AERODYNAMIC PRINCIPLES AND CONSEQUENCES

It is the first book in a large and special series of books, dedicated to motorsport in general; it will cover aerodynamics, suspension, engines, dynamics, etc. ..... Everything you need to learn how to design a full car.

The aim of this series is also to say that I would like to teach again in a university.

I hope that this series will be a success and that I will be able to transmit all my knowledge and all my experience.

@TimoteoBriet

Let's look at the most important principles and consequences of air aerodynamics.

BERNOULLI'S PRINCIPLE

The Bernoulli's principle, is applicable to stationary ideal fluids with constant density. An ideal fluid is incompressible and only superficial forces are acting on it due to pressure. There is no mass transport, heat or momentum. Therefore, they are inviscid fluids (No viscosity). Moreover, a fluid is stationary if its variables' partial derivates with respect to time are null. This means that these variables are constant with respect to time in any point in space, although, obviously, they can vary from point to point.

From this principle, we can say that the energy is the same at any point of a current line of a fluid (trajectory tangent to the velocity field at any point). Let's think of an air flow that flows along a duct. We establish two sections with areas "S_1" and "S_2", which are crossed by this flow and are separated by a differential of length "dl". "V" indicates the volume and "v" indicates the fluid velocity. The forces on the surfaces "S_1" y "S_2" are:

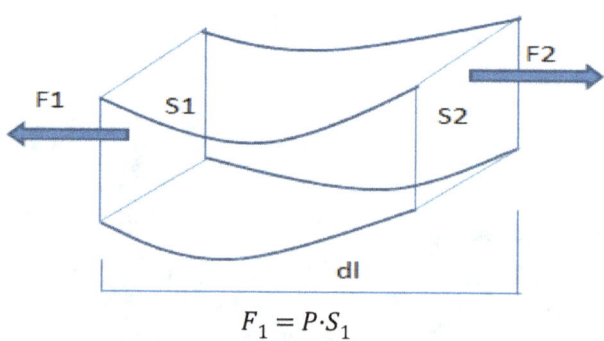

$$F_1 = P \cdot S_1$$
$$F_2 = -(P + dP) \cdot S_2$$

We have considered the pressure direction to be opposite to the normal to each surface between which there is a difference of pressure dP. The addition of both forces is determined by:

$$F_T = F_1 + F_2$$

The surfaces are separated by a "dl". Therefore, we know the following: $S_1 = S_2 = S$

$$F_T = P \cdot S - (P + dP) \cdot S = - S \cdot dP$$

$$F_T = F_1 + F_2$$

The acceleration will be determined by:

$$a = \frac{dv}{dt}$$

And the mass:

$$m = \rho \cdot dV = \rho \cdot S \cdot dl$$

With Newton's second law F= m·a, and knowing v= dl/dt:

$$- S \cdot dP = \rho \cdot S \cdot dl \cdot \frac{dv}{dt}$$
$$- S \cdot dP = \rho \cdot S \cdot v \cdot dv$$
$$\rho \cdot v \cdot dv + dP = 0$$

This is the Bernoulli's principles in differential form.

If the March number is always lower than 0.2, we can assume constant density. Integrating:

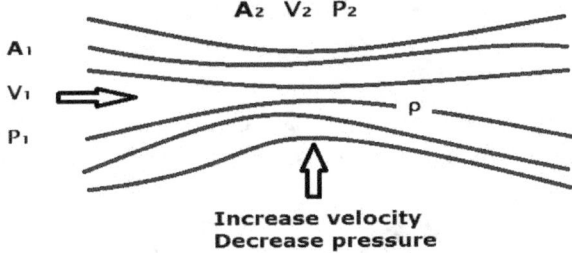

A₂ V₂ P₂

A₁

V₁ ⇨

P₁

P

⇧
Increase velocity
Decrease pressure

$$P + \frac{1}{2} \cdot \rho \cdot v^2 = cte$$

"P" is the dynamic pressure.
Energy per unit volume before = Energy per unit volume after

$$P_1 + \frac{1}{2} \cdot \rho \cdot v_1{}^2 + \rho \cdot g \cdot h_1 = P_2 + \frac{1}{2} \cdot \rho \cdot v_2{}^2 + \rho \cdot g \cdot h_2$$

With $A_1 > A_2$, by continuity $v_1 < v_2$ and by the Bernoulli's principle $P_1 > P_2$

The next 2 pictures, have the same effect (high speed, with high pressure):

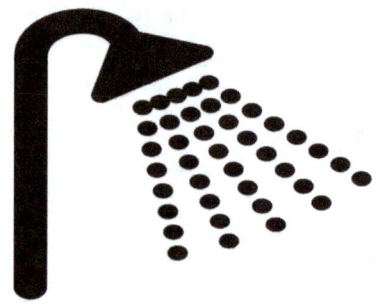

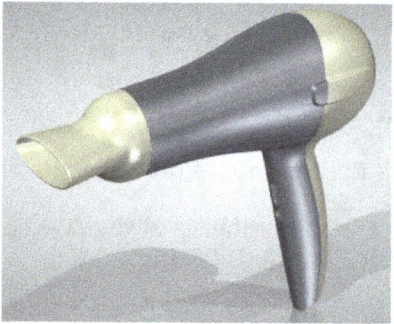

If the holes are big, is necessary to do that entry a lot water (mass rate); if the holes are very big, in this example we can "see" an exit rocket iiii:

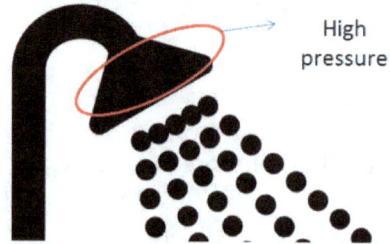

High pressure

The Bernoulli's principle, the addition of the kinetic energy, potential energy and energy associated with the pressure remains constant when we move along a straight line. As the flow must be constant (incompressible fluid, as we have said), if a duct reduces the section of fluid, its velocity will increase. From Bernoulli's point of view: if the height of both points is the same, the velocity will increase if the pressure decreases.

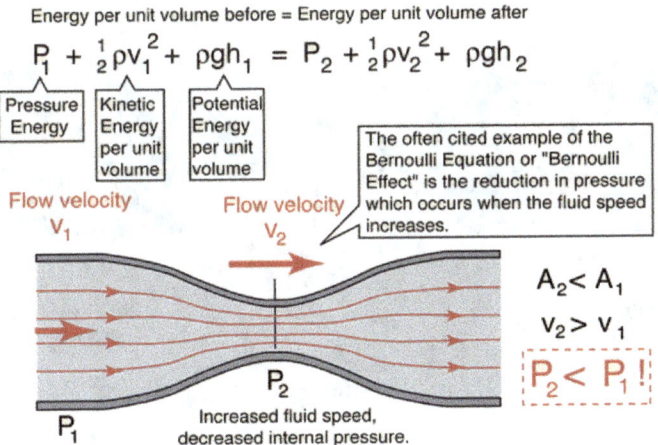

Energy per unit volume before = Energy per unit volume after

$$P_1 + \frac{1}{2}\rho v_1^2 + \rho g h_1 = P_2 + \frac{1}{2}\rho v_2^2 + \rho g h_2$$

| Pressure Energy | Kinetic Energy per unit volume | Potential Energy per unit volume |

The often cited example of the Bernoulli Equation or "Bernoulli Effect" is the reduction in pressure which occurs when the fluid speed increases.

Flow velocity V_1

Flow velocity V_2

$A_2 < A_1$

$V_2 > V_1$

$P_2 < P_1$!

P_2

P_1

Increased fluid speed, decreased internal pressure.

If the section decreases in a duct (without loss), the fluid must move faster, because the same quantity must travel through a smaller section ($Q = A \cdot v = $ cte).

As the addition of the three energies of Bernoulli is always constant (we suppose that the height is the same, taking two points (wide section and narrow section) at the same height), the pressure must increase, because if the flow is constant, the velocity increases. What we have deduced is exactly what the equations say. All we deduce must have a mathematical background and also have to be quantified. This is the fundamental work of mathematicians and physicists: numerical modeling of reality.

Chimney:

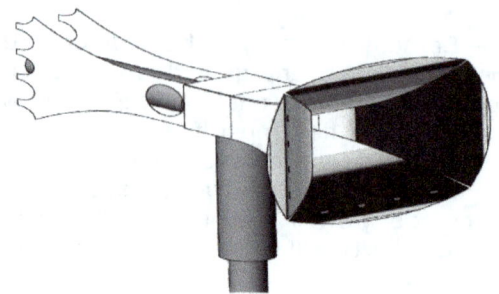

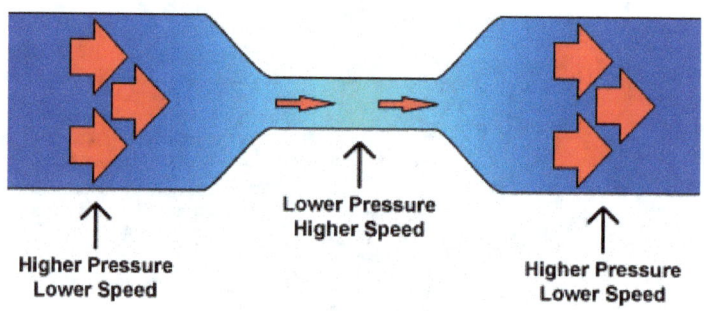

Lower Pressure
Higher Speed

Higher Pressure
Lower Speed

Higher Pressure
Lower Speed

Let's look at a deduction of the Bernoulli's principle:

Considering the mass conservation, we have the following:

$$\frac{d}{dt}\delta m = \frac{d}{dt}(\rho \cdot \delta V) = 0$$

Where we have use the relation:

$$\delta m = \rho \cdot \delta V$$

Developing the first expression:

$$\frac{d\rho}{dt} \cdot \delta V + \rho \cdot \frac{d(\delta V)}{dt} = \frac{d\rho}{dt} \cdot \delta V + \rho \cdot \nabla \cdot \bar{v} \, \delta V = 0$$

The continuity equation can be described by the Euler equation:

$$\frac{\partial \rho}{\partial t} + \nabla \cdot (\rho \cdot \vec{v}) = 0$$

or:

$$\frac{\partial \rho}{\partial t} + \vec{v} \cdot \nabla \cdot \rho + \rho \cdot \nabla \cdot \vec{v} = 0$$

The gravity is an external and conservative force of volume that is independent of time. As it is a conservative force, we can, using the breakdown of Kelvin-Helmholtz, write it as the gradient of a scalar function (called gravitational potential) as follows:

$$\vec{f} = \rho \cdot \text{g} \cdot \hat{z} = \rho \cdot \overline{\nabla \emptyset}$$

$$\vec{v} \cdot \vec{f} \cdot \delta V = \vec{v} \cdot \rho \cdot \overline{\nabla \emptyset} \cdot \delta V = \left(\frac{d\emptyset}{dt} - \frac{\partial \emptyset}{\partial t}\right) \cdot \rho \cdot \delta V = \frac{d}{dt}(\emptyset \cdot \rho \cdot \delta V) - \rho \cdot \frac{\partial \emptyset}{\partial t} \cdot \delta V$$

Considering that the fluid is not changing, we can deduce the following:

$$\frac{d}{dt}\left(\frac{1}{2} \cdot \rho \cdot v^2 \cdot \delta V + p \cdot \delta V - \emptyset \cdot \rho \cdot \delta V + u \cdot \rho \cdot \delta V\right) = \frac{\partial p}{\partial t} \cdot \delta V - \rho \cdot \frac{\partial \emptyset}{\partial t} \cdot \delta V$$

$$\frac{d}{dt}\left(\frac{1}{2} \cdot v^2 + \frac{p}{\rho} - \emptyset + u\right) = \frac{1}{\rho} \cdot \frac{\partial p}{\partial t} - \frac{\partial \emptyset}{\partial t}$$

With it, we arrive to the following expression:

$$\frac{1}{2} \cdot v^2 + \frac{p}{\rho} + g \cdot z + u = k = \text{cte}$$

We must remember the properties of the fluid needed to reach Bernoulli's principle expression:

- Incompressible fluid
- Inviscid fluid
- Irrotational fluid

Besides considering these properties, we have used the gravity as a conservative force of volume and the Navier-Stokess equations.

$$\frac{1}{\rho} \nabla P = \nabla \frac{p}{\rho}$$

$$\mu = 0$$

$$\nabla \wedge \vec{v} = 0 \rightarrow \vec{v} = \nabla \phi$$

$$\vec{f} = -\rho g z$$

$$\frac{\partial \vec{v}}{\partial t} = \frac{\partial}{\partial t} \nabla \phi = \nabla \frac{\partial \phi}{\partial t}$$

So:

$$\frac{\partial \vec{v}}{\partial t} + \left(\vec{v} \vec{\nabla} \right) \vec{v} - \frac{\vec{f}}{\rho} + \frac{1}{\rho} \nabla P = 0$$

$$\nabla \frac{\partial \phi}{\partial t} + \nabla \frac{v^2}{2} - \nabla \left(- gz \right) + \nabla \frac{p}{\rho} = 0$$

$$\frac{\partial \phi}{\partial t} + \frac{v^2}{2} + gz + \frac{p}{\rho} = c(t)$$

This is a fundamental principle needed in car design, as an approximation. The two premises (sometimes consequences) we have to work with are:

- If the area is reduced, the velocity increases and the pressure decrease.
- If the area increases, the velocity decreases and the pressure increase.

There are many "tricks" to observe and validate Bernoulli: let's take, for example, two cans of coke and blow through the space between them. The cans will tend to join each other due to the low pressure that exists between them:

Another possibility of this principle is to know flow parameters (velocity and pressure), depending on known parameters of the flow in other zones.

Let's imagine a converging duct through which air flows. Let's suppose that we work in two dimensions and the horizontal axis is "x" and the vertical axis is "y". Its directional velocities are "u" and "v". "A, B, C, D" are four points in the central part of the duct (assuming constant density: incompressible fluid):

$$a_x = u\frac{\partial u}{\partial x} + v\frac{\partial u}{\partial y}$$

$$\frac{\partial v}{\partial y} = -\frac{\partial u}{\partial x}$$

Therefore, we can obtain the velocities in all points of the duct, knowing the velocities of the aforementioned four points. Obviously, a smooth passage section does not imply a "uniform" distribution of the fluid along the duct.

Applying Bernoulli implies this assumption, wrong! Considering the viscosity is something essential. When analyzing a duct with section change, "only applying Bernoulli", we can make many mistakes. The most common one is to think that the flow fills the pipe where there's a change in section:
Density is constant: → mistake!

The density can also change, not only the velocity!

For example, let's consider a duct which changes suddenly of section:

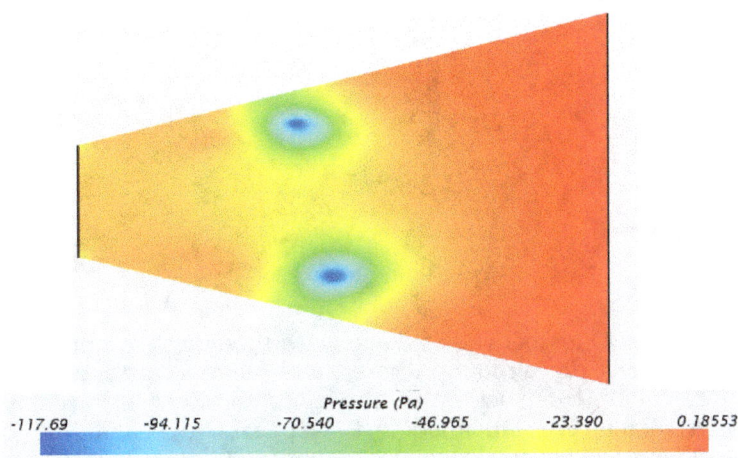

Pressure (Pa)

| -117.69 | -94.115 | -70.540 | -46.965 | -23.390 | 0.18553 |

That is the reality; there are two vortex and so, a very important concept:

It's impossible to know the speed and pressure distribution in a cannel using ONLY Bernouilli:

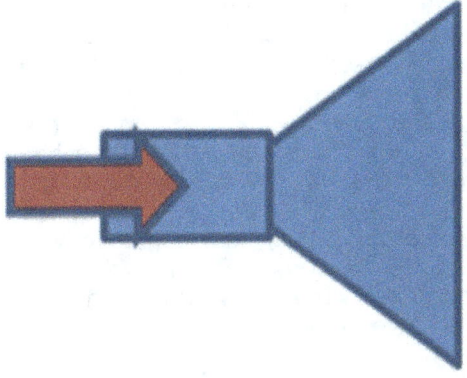

Considering that all the flow entering the section will fill all the pipe is something common to assume. This isn't always true. However, when considering only Bernoulli, we are supposing it. The reality is different and we can observe it in many situations and contexts:

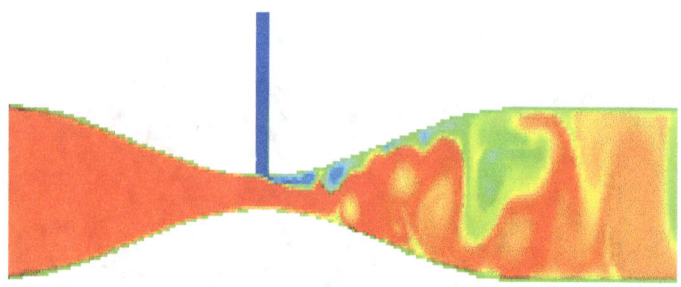

One of the things we shouldn't forget and shouldn't do (although it is easy to make a mistake) is to calculate the suction of the ground effect using only Bernoulli. We can imagine the space between the car's floor and the pavement to be like a tube or duct where the section varies. Therefore, we can use Bernoulli and the energy conservation to calculate the existent depression depending on the velocity and, therefore, the generated *downforce*. We would definitely make a big mistake because in ground effect, the viscosity is the most influential parameter and Bernoulli doesn't consider it. We also have to think that it isn't a real duct, because it isn't perfectly closed: there is air flow that goes in and out through the sides.

Regarding the design of car under trays, the use of excessive angles may lead to the detachment of the boundary layer, thus we must restrict its value to 10-15º.

The floor of racing cars can take very different shapes, but they always seek to reach the highest possible depression and thus, the highest *downforce*. This fact is achieved modifying the section of airflow:

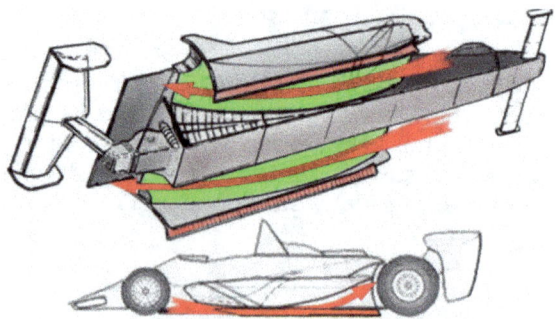

In the following chapters we will cover, racing car diffusers, as a direct application of Bernoulli. The airflow section increases to lower velocity and increases pressure (which is lower than the external pressure), generating *downforce*:

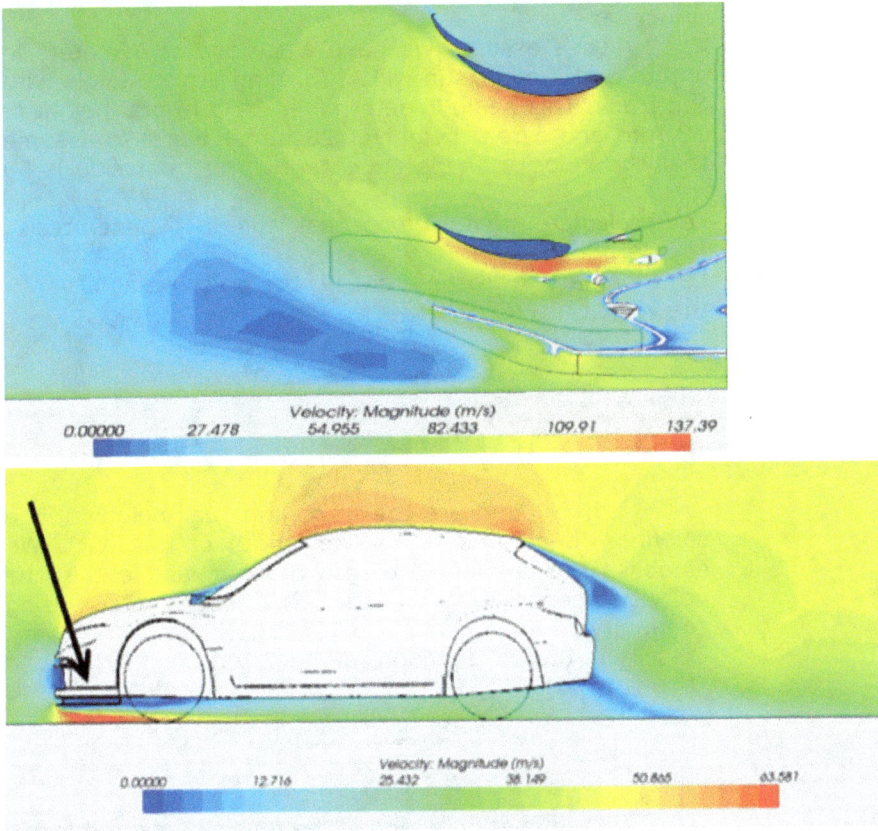

When we try to apply the Bernoulli's principle to the ground:

- We are omitting the existence of the viscosity: the viscosity causes effects of compressibility and plugs, stopping the airflow below the floor. Also, as we have previously said, the detachment of the boundary layer will damage its functioning.
- We are also omitting the absence of side walls. As a result there is air that "wants" to enter the low pressure zone, destroying

the created *downforce*.

Let's look at a specific example and apply an energy balance to check the difference in value between the *downforce* calculated using 'CFD' software and the *downforce* which is calculated by applying the Bernoulli's principle:

Let's assume we have a surface of 4 meters in length and 2 meters in width, forming a pitch angle with the track. This angle is determined by a frontal height of 10 mm and a rear height of 20 mm. Let's also assume that the velocity of the air entering the front side is 50 m/s.

Let's calculate the *downforce* that "the duct" generates (by Bernoulli):

Let's look at: "A" is the area, "P" the pressure and "v" the velocity:

$$v_{intake} \cdot A_{intake} = v_{exhaust} \cdot A_{exhaust}$$

$$\rho \cdot \frac{v_{intake}^2}{2} + P_{intake} = \rho \cdot \frac{v_{exhaust}^2}{2} + P_{exhaust}$$

"h_1" y "h_2" are the front and rear height of the car-floor "duct" and "L" is the length of this duct. We establish the conditions of the problem and obtain the velocity depending on the "x" coordinate, using continuity:

Velocity in the "x" direction will be determined by:

$$v_x = \frac{A_{intake}}{A_x} \cdot v_\infty$$

If we specify, we have the following expression:

$$v_x = \frac{h_1}{h_1 + \frac{(h_2 - h_1)}{L} \cdot x} \cdot v_\infty$$

Let's notice that $v_{intake} \equiv v_\infty$. We apply Bernoulli between 0 and 1 and obtain the gauge pressure expression depending on x:

$$P_\infty = \frac{\rho}{2} \cdot v_\infty^2 \cdot \left(1 - \frac{h_1}{\left(h_1 + \dfrac{h_2 - h_1}{L} \cdot x\right)^2}\right)$$

We integrate the gauge pressure of the whole diffuser surface and then, we obtain the generated *downforce*:

$$F = b \cdot \int P_{max}(x) \cdot dx$$

We obtain from the numerical data considered at the beginning of this example (h_1=10mm, h_2=20mm, L=4m y b=2m), more or less, 3836, 25 N of *downforce*. This result is much higher than what we would have obtained if we had the same conditions in a wind tunnel or at a racetrack.

Let's look at a more realistic way of calculation using CFD:

Let's use the same dimensions (4 x 2 meters) with the same height with respect to the track and the same pitch angle. Equally, we maintain the same car velocity with respect to the air (50m/s).

The CFD simulation will only give us the generated *downforce* on the lower surface of the car, on the floor. Doing this is only possible with CFD. We cannot do this with wind tunnels, although they are big, expensive or even real. This is one of the greatest advantages of CFD. Besides, we will observe the air dynamics around and under the floor and, this is always helpful.

Once we run the simulation, the obtained *downforce* value is 792 N. Comparing this value with the one obtained using the Bernoulli (3836 N), we can say the following:

Supposing the air "doesn't try to get in" the generated depression on the ground chamber, and also supposing that all air which goes in, leaves the chamber uniformly, or not considering the viscosity, means that we are making a mistake four times the obtained value by CFD. The *downforce* is five times lower! It's incredible but it is true.

From here, we can deduce the importance of placing a diffuser on the back of the car with the goal that we know. Generating vortexes along the floor is also important, sealing the floor's depression from air at ambient pressure.

With all this, we can conclude saying that the highest *downforce* that we can obtain on the car's floor is the quantity obtained using Bernoulli. It is a limit that we cannot surpass. Another limit for that is balance energy for deflection air (Speed and mass).

Another limit for that is may be, more important; that is; before object to study his downforce, the air have a speed and mass; that two parameters produce one movement quantity; that is: one energy; this energy is the maximum that the downforce can reached iiii.

On the other hand, 792 N is not a negligible quantity. The use of a diffuser would be equal to increasing the *downforce* a lot...

We can try to simulate the Bernoulli flux in floor and diffuser.

Step 1: Fix velocity inlet.

Setp 2: Fix velocity outlet.

ONLY FLOOR

STEP 1:

Height front (h_f)	10 mm
Wingspan (b)	1,8 m
Total long (l_m)	3,5 m
Velocity (V_{inf})	60 m/s
Height rear (h_r)	Between 10mm and 100mm

$$V(x) = \frac{A_i V_{inf}}{A(x)} \quad , \quad A(x) = b\left(h_f + x\frac{(h_r - h_f)}{x_m}\right)$$

$$P(x) = 0,5 * 1,225 * (V_r^2 - V(X)^2) \quad , \quad Vr = V(x = x_m)$$

Integrating between 0 and x_m, we obtain downforce:

$$DF = b * \cos(\alpha) * 0,5 * 1,225 \left[x_m V_r^2 - \frac{(A_i V_{inf})^2}{b^2 * \tan(\alpha)}\left(\frac{1}{h_f} - \frac{1}{h_f + x_m \tan(\alpha)}\right)\right],$$

$$\alpha = \arctan\left(\frac{(h_r - h_f)}{x_m}\right)$$

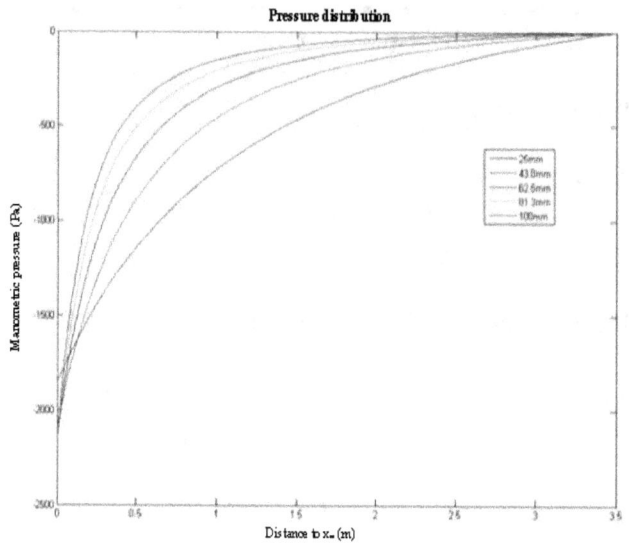

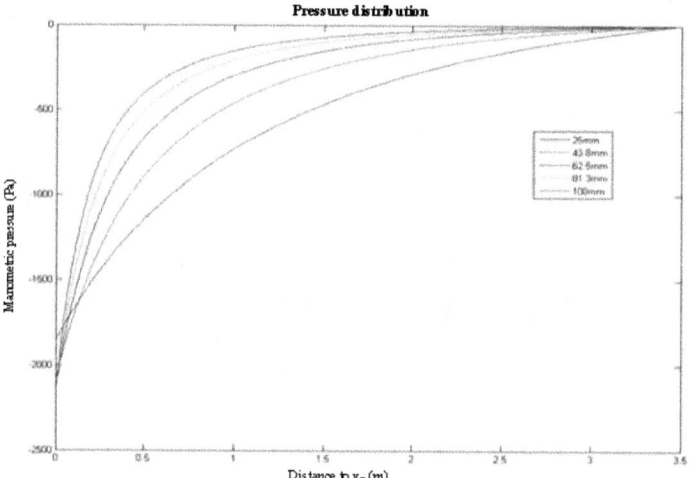

Total downforce:

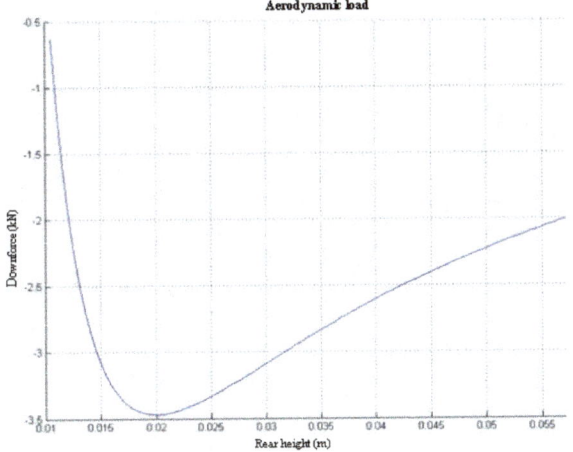

STEP 2:

$$V(x) = \frac{A_r V_{inf}}{A(x)} \quad , \quad A(x) = b\left(h_f + x\frac{(h_r - h_f)}{x_m}\right)$$

$$P(x) = 0.5 * 1.225 * \left(V_{inf}^2 - V(X)^2\right)$$

Integrating:

$$DF = b * \cos(\alpha) * 0.5 * 1.225 \left[x_m V_{inf}^2 - \frac{\left(A_r V_{inf}\right)^2}{b^2 * \tan(\alpha)}\left(\frac{1}{h_f} - \frac{1}{h_f + x_m \tan(\alpha)}\right)\right],$$

$$\alpha = \arctan\left(\frac{(h_r - h_f)}{x_m}\right)$$

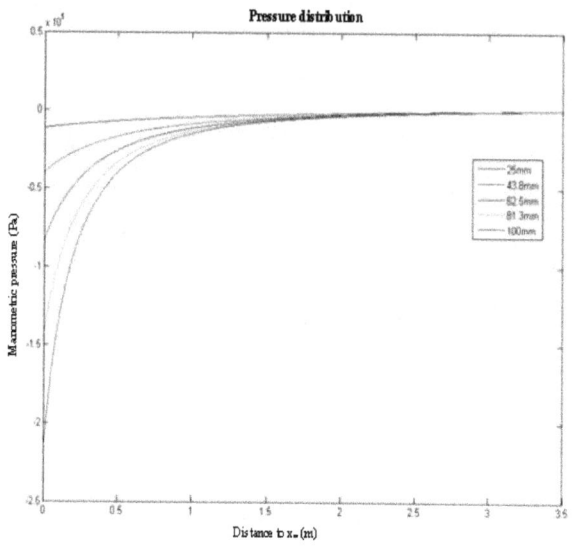

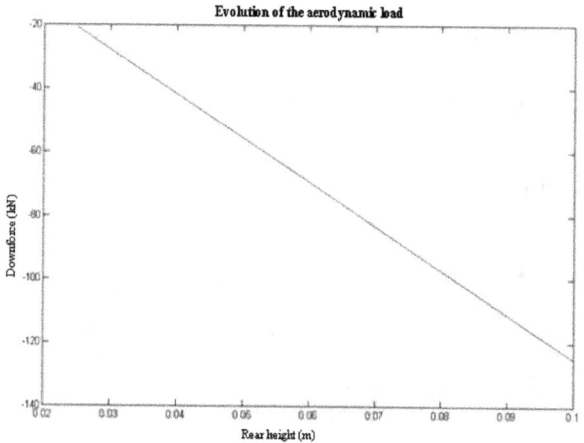

FLOOR AND DIFUSSER:

STEP 1:

Height front (h_f)	10 mm
Wingspan (b)	1,8 m
Total long (l_m)	3,5 m
Velocity (V_{inf})	60 m/s
Height rear (h_r)	Between 10mm and 100mm
Long floor (l_s)	Between 2 and 3 m
Long difussor (l_d)	l_m l_s
Difussor angle (a_d)	Between 10 and 30 degrees

$$V_d(x) = \frac{A_i V_{inf}}{A_d(x)} \quad , \quad A_d(x) = b(h_r + (x - l_s)\tan(\alpha_d)),$$

$$\forall x \mid l_s \leq x \leq l_m$$

$$P_d(x) = 0,5 * 1,225 * (V_r^2 - V_d(X)^2), V_r = V_d(x = l_m),$$
$$\forall x \mid l_s \leq x \leq l_m$$

$$V_s(x) = \frac{A_i V_{inf}}{A_s(x)}, \quad A_s(x) = b\left(h_f + x\frac{(h_r - h_f)}{l_s}\right), \quad \forall x \leq l_s$$

$$P_s(x) = 0,5 * 1,225 * (V_r^2 - V_s(X)^2) \, \forall x \leq l_s$$

Integrating:

$$DF_s = b * \cos(\alpha_s) * 0.5 * 1.225 \left[l_s V_r^2 - \frac{(A_i V_{inf})^2}{b^2 * \tan(\alpha_s)} \left(\frac{1}{h_f} - \frac{1}{h_f + l_s \tan(\alpha_s)} \right) \right],$$

$$\alpha_s = \arctan\left(\frac{(h_r - h_f)}{l_s} \right)$$

$$DF_d = b * \cos(\alpha_d) * 0.5 * 1.225 \left[l_d V_r^2 - \frac{(A_i V_{inf})^2}{b^2 * \tan(\alpha_d)} \left(\frac{1}{h_r} - \frac{1}{h_r + l_d \tan(\alpha_d)} \right) \right],$$

$$DF = DF_s + DF_d$$

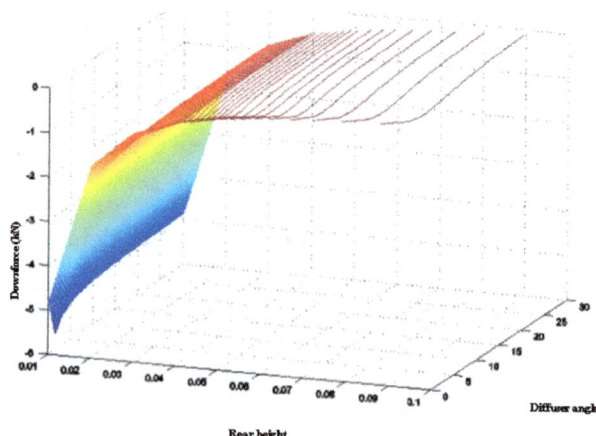

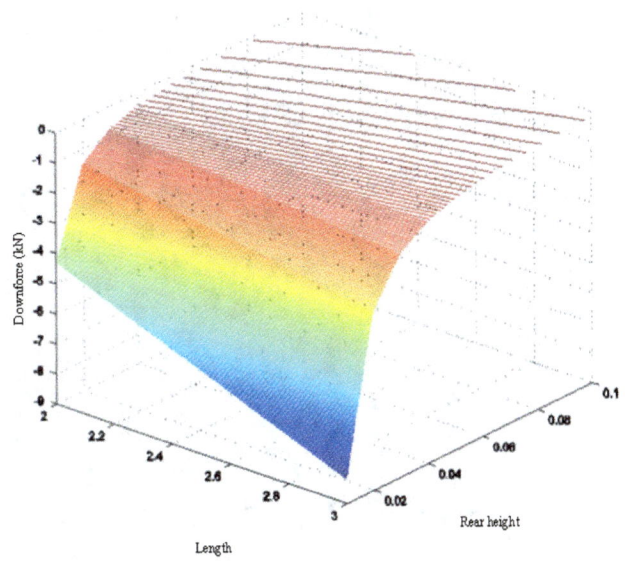

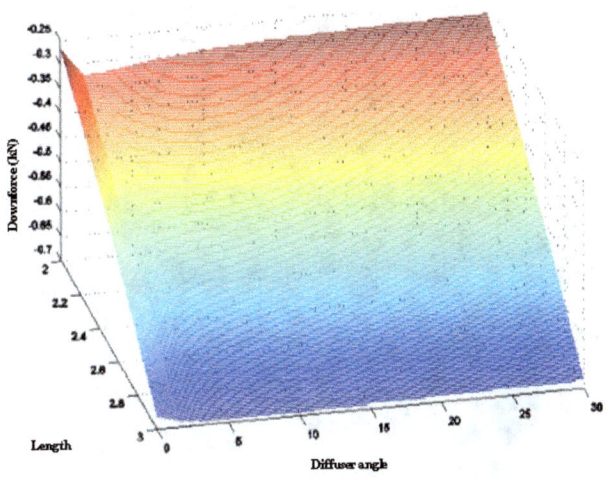

STEP 2:

$$V_d(x) = \frac{A_r V_{inf}}{A_d(x)}, A_d(x) = b(h_r + (x - l_s)\tan(\alpha_d)), \qquad \forall x \mid l_s \leq x \leq l_m$$

$$P_d(x) = 0{,}5 * 1{,}225 * \left(V_{inf}^2 - V_d(X)^2\right) \forall x \mid l_s \leq x \leq l_m$$

$$V_s(x) = \frac{A_r V_{inf}}{A_s(x)}, \quad A_s(x) = b\left(h_f + x\frac{(h_r - h_f)}{l_s}\right), \qquad \forall x \leq l_s$$

$$P_s(x) = 0{,}5 * 1{,}225 * \left(V_{inf}^2 - V_s(X)^2\right) \forall x \leq l_s$$

Integrating:

$$DF_s = b * \cos(\alpha_s) * 0{,}5 * 1{,}225 \left[l_s V_{inf}^2 - \frac{(A_r V_{inf})^2}{b^2 * \tan(\alpha_s)}\left(\frac{1}{h_f} - \frac{1}{h_f + l_s \tan(\alpha_s)}\right)\right],$$

$$\alpha_s = \arctan\left(\frac{(h_r - h_f)}{l_s}\right)$$

$$DF_d = b * \cos(\alpha_d) * 0{,}5 * 1{,}225 \left[l_d V_{inf}^2 - \frac{(A_r V_{inf})^2}{b^2 * \tan(\alpha_d)}\left(\frac{1}{h_r} - \frac{1}{h_r + l_d \tan(\alpha_d)}\right)\right],$$

$$DF = DF_s + DF_d$$

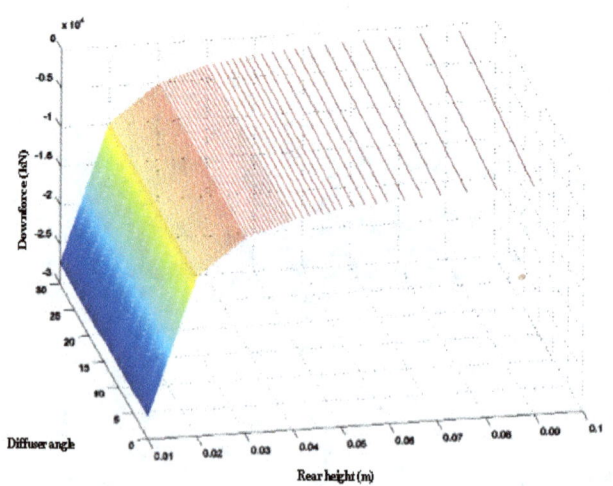

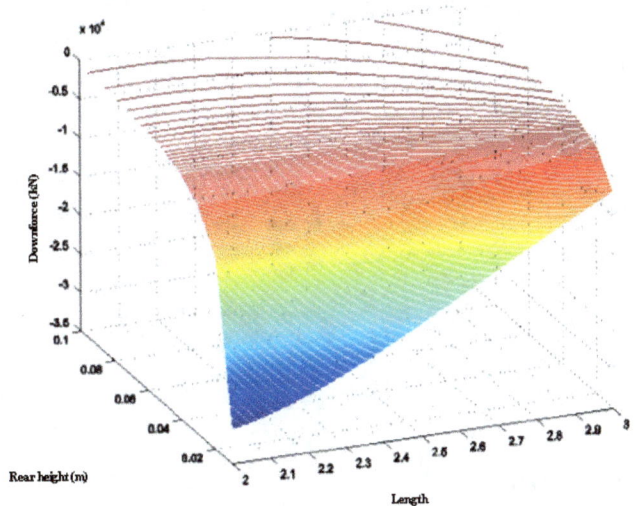

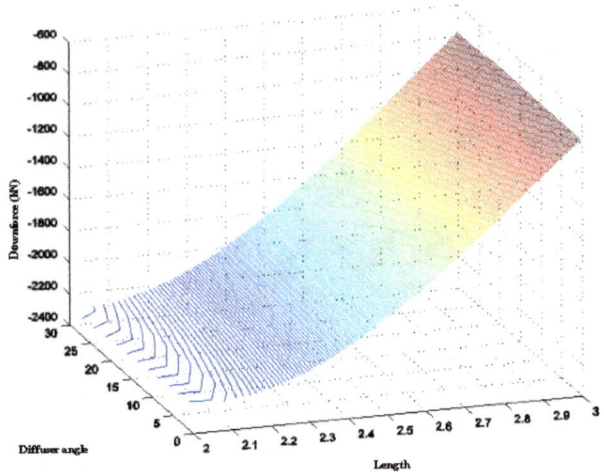

The matlab code used to plot the last graphs, is:
Floor only, velocity fix inlet:
Tunnel

```
clear all
close all
clc
n=1000
hf=10e-3;
hr=linspace(10e-3, 100e-3,n);
%hr=64e-3;
xm=3.5;
b=1.8;
Vinf=60;
a=atan((hr-hf)/xm);
Ai=hf*b;
Ar=b.*hr;
Vr=(Ai*Vinf)./Ar;
Ci=Ai*Vinf;
Ci2=0.5*1.225.*Vr.^2;
figure
hold on
Pcheck=[];
vcheck=[];
for i=1:n
x=linspace(0, 3.5, n);
V=Ci./(b.*(hf+x.*tan(a(i))));
P=Ci2(i)-0.5*1.225.*V.^2;
Pcheck=[Pcheck; P];
vcheck=[vcheck; V];

end

plot(x, Pcheck(1:200:end,:))
```

Downforce

```
DF=[];
for j=1:n
    a=atan((hr(j)-hf)/xm);
    df=0.001*(b*cos(a))*trapz(x,
Pcheck(j,:));
    DF=[DF df];
    clear df
end
figure
plot(hr, DF) DF25=spline(hr, DF, 20e-3)

DF25=spline(hr, DF, 20e-3)
```

```matlab
clear; clc; close all;
n=1000;
hr=linspace(10e-3, 100e-3, n);
x=linspace(0,3.5,1000);
hf=10e-3
xm=3.5
ro=1.225;
Vinf=60;
b=1.8;
Ai=b*hf;
Cr2=0.5*ro*Vinf^2;
Ps=[];
for i=1:n
    a=atan((hr(i)-hf)/xm);
    A=b.*(hf+x.*tan(a));
    Ar=A(end);
    Cr=Ar*Vinf;
    V=Cr./A;
    P=Cr2-(0.5*ro.*V.^2);
    Ps=[Ps;P];
    clear P
end
plot(x, Ps(1:200:end, :))

DF=[];
for j=1:n
    a=atan((hr(j)-hf)/xm);
    df=0.001*(b*cos(a))*trapz(x, Ps(j,:));
    DF=[DF df];
    clear df
end
% plot(x, Ps)
%     legend('25mm', '43.8mm', '62.5mm',
'81.3mm', '100mm')
% xlabel('distancia al extremo de la placa (m)')
% ylabel('Presion manométrica (Pa)')
% title('Distribución de presiones')
figure
plot(hr, DF)

%%DF 25mm  en kN
DF25= spline(hr, DF, 25e-3);
```

Floor and diffusor, velocity fix front:

nss

```
ns=20;
nd=20;
nr=10;
```

Geometry

```
x=linspace(1, 3.5, 2500);
hf=10e-3; % m
b=1.8; %m
lm=3.5; %m
Vinf= 60; %m/s
ls=linspace(2, 3, ns); %m
ld=lm-ls; %m
ad=linspace(0, 30, nd); %deg
hr=linspace(10e-3, 100e-3, nr); %m
Ai=b*hf; %m2
Ci=Vinf*Ai;
    Ps=zeros(2500, ns, nr, nd);
```

```matlab
for k=1:ns
    temp=find(abs(x-ls(k))<5e-3, 1,'last');
    xs=x(1:temp);
    xd=x(temp+1:end);


    for i=1:nr
        hxs=hf+xs.*((hr(i)-hf)/xs(end));
        Axs=b*hxs;
        Vxs=Ci./Axs;
%        Ar=Axs(end);
%        Vr=Vxs(end);


        for j=1:nd
            hxd=(hr(i)+(xd-xs(end)).*tand(ad(j)));
            Axd=b.*hxd;
            Ad=Axd(end);
            Vxd=Ci./Axd;
            Vd=Vxd(end);
            Pxd=0.5*1.225.*(Vd^2-Vxd.^2);
            Pxr=0.5*1.225*(Vd^2-Vxs.^2);
            P=[Pxr Pxd];
            Ps(:,k,i,j)=P;   %Pa
            clear P;
        end
    end
end
```

```matlab
%DF computation
DF=zeros(ns, nr,nd);
for k=1:ns
    temp=find(abs(x-ls(k))<5e-3, 1,'last');
    xs=x(1:temp);
    xd=x(temp+1:end);
    for i=1:nr
        as=atan((hr(i)-hf)/ls(k));
        DFs=0.001*(b*cos(as))*(trapz(xs, Ps(1:temp,k,i,j)));
        for j=1:nd

            DFd=0.001*(b*cosd(ad(j)))*(trapz(xd, Ps(temp+1:end, k,i,j)));
            Df=DFs+DFd;
            DF(k,i,j)=Df;
        end
    end
end


temp2=[5];


for i=1:nr
    for j=1:nd
        Z1(i,j)=DF(temp2,i,j);
    end
end
for k=1:ns
    for i=1:nr
        Z2(k,i)=DF(k,i,temp2);
    end
end
for k=1:ns
    for j=1:nd
        Z3(k,j)=DF(k,temp2,j);
    end
end
figure
contour3(hr, ad, Z1', 200)
xlabel('altura del fondo')
ylabel('angulo difusor')
zlabel('Downforce')
figure
contour3(ls, hr, Z2', 200)
ylabel('altura del fondo')
xlabel('longitud fondo')
zlabel('Downforce')
figure
contour3(ls, ad, Z3', 200)
xlabel('longitud del fondo')
ylabel('angulo difusor')
zlabel('Downforce')
```

Floor and diffusor, velocity fix outlet:

Nss

```
ns=20;
nd=20;
nr=10;
```

Geometry

```
x=linspace(1, 3.5, 2500);
hf=10e-3; % m
b=1.8; %m
lm=3.5; %m
vinf= 60; %m/s
ls=linspace(2, 3, ns); %m
ld=lm-ls; %m
ad=linspace(0, 30, nd); %deg
hr=linspace(10e-3, 100e-3, nr); %m

 Ps=zeros(2500, ns, nr, nd);
```

```
for k=1:ns
    temp=find(abs(x-ls(k))<5e-3, 1,'last');
    xs=x(1:temp);
    xd=x(temp+1:end);


    for i=1:nr
       hxs=hf+xs.*((hr(i)-hf)/xs(end));
       Axs=b*hxs;

%       Ar=Axs(end);
%       Vr=Vxs(end);


        for j=1:nd
            hxd=(hr(i)+(xd-xs(end)).*tand(ad(j)));
            Axd=b.*hxd;
            Ad=Axd(end);
            Cd=Vinf*Ad;
            Vxd=Cd./Axd;
            Vxs=Cd./Axs;
            Vd=Vxd(end);
            Pxd=0.5*1.225.*(Vd^2-Vxd.^2);
            Pxr=0.5*1.225*(Vd^2-Vxs.^2);
            P=[Pxr Pxd];
            Ps(:,k,i,j)=P;   %Pa
            clear P;
        end
    end
end
```

```matlab
%DF computation
DF=zeros(ns, nr,nd);
for k=1:ns
    temp=find(abs(x-ls(k))<5e-3, 1,'last');
    xs=x(1:temp);
    xd=x(temp+1:end);
    for i=1:nr
        as=atan((hr(i)-hf)/ls(k));
        DFs=0.001*(b*cos(as))*(trapz(xs, Ps(1:temp,k,i,j)));
        for j=1:nd

            DFd=0.001*(b*cosd(ad(j)))*(trapz(xd, Ps(temp+1:end,
k,i,j)));
            Df=DFs+DFd;
            DF(k,i,j)=Df;
        end
    end
end


temp2=[5];


for i=1:nr
    for j=1:nd
        Z1(i,j)=DF(temp2,i,j);
    end
end
for k=1:ns
    for i=1:nr
        Z2(k,i)=DF(k,i,temp2);
    end
end
for k=1:ns
    for j=1:nd
        Z3(k,j)=DF(k,temp2,j);
    end
end
figure
contour3(hr, ad, Z1', 200)
xlabel('altura del fondo')
ylabel('angulo difusor')
zlabel('Downforce')
figure
contour3(ls, hr, Z2', 200)
ylabel('altura del fondo')
xlabel('longitud fondo')
zlabel('Downforce')
figure
contour3(ls, ad, Z3', 200)
xlabel('longitud del fondo')
ylabel('angulo difusor')
zlabel('Downforce')
```

Another example: Calculate the total downforce in this undertray and diffuser set.

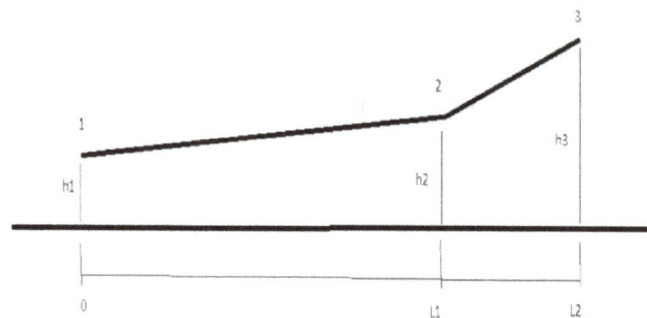

We know the downforce is equal to:

$$F = b \int_0^{L_2} P(x)\,dx \quad (1)$$

Calculate the pressure in **0<x<L₁**, using Bernouilli:

$$P(x) = \frac{1}{2}\rho(v_3^2 - v_x^2) \quad (2)$$

Using continuity and we assuming that the velocity is the velocity in point (1):

$$v_3 \cdot b \cdot h_3 = v_1 \cdot b \cdot h_1 \rightarrow v_3 = \frac{h_1}{h_3}v_\infty \quad (3)$$

$$v_x \cdot b \cdot h_x = v_1 \cdot b \cdot h_1 \rightarrow v_x = \frac{h_1}{h_x}v_\infty \quad (4)$$

So:

$$h_x = h_1 + x \cdot \tan(\alpha_1) \quad (5)$$

Where:

$$\tan(\alpha_1) = \frac{h_2 - h_1}{L_1} \quad (6)$$

Substituting (3) and (4) in (2):

$$P(x) = \frac{1}{2}\rho v_\infty^2 \left[\left(\frac{h_1}{h_3}\right)^2 - \frac{h_1^2}{(h_1 + x\tan(\alpha_1))^2}\right] \quad (7)$$

If the density and velocity are constants in all the length, the force is:

$$F_1 = \frac{1}{2}\rho v_\infty^2 \int_0^{L_1} \left[\left(\frac{h_1}{h_3}\right)^2 - \frac{h_1^2}{(h_1 + x\tan(\alpha_1))^2}\right] dx$$

$$= \frac{1}{2}\rho v_\infty^2 \left\{\left(\frac{h_1}{h_3}\right)^2 L_1 + \frac{h_1^2}{\tan(\alpha_1)}\left[\frac{1}{h_1 + L_1\tan(\alpha_1)} - \frac{1}{h_1}\right]\right\} \quad (8)$$

Now, the second length: **$L_1 < x < L_2$**. In this case:

$$h_x = h_2 + (x - L_1) \cdot \tan(\alpha_2) \quad (9)$$

Where:

$$\tan(\alpha_2) = \frac{h_3 - h_2}{L_2 - L_1} \quad (10)$$

Substituting in (2):

$$P(x) = \frac{1}{2}\rho v_\infty^2 \left[\left(\frac{h_1}{h_3}\right)^2 - \frac{h_1^2}{(h_2 + (x - L_1)\tan(\alpha_2))^2}\right] \quad (11)$$

$$F_2 = \frac{1}{2}\rho v_\infty^2 \int_{L_1}^{L_2} \left[\left(\frac{h_1}{h_3}\right)^2 - \frac{h_1^2}{(h_2 + (x - L_1)\tan(\alpha_2))^2}\right] dx$$

$$= \frac{1}{2}\rho v_\infty^2 \left\{\left(\frac{h_1}{h_3}\right)^2 (L_2 - L_1)\right.$$

$$\left. + \frac{h_1^2}{\tan(\alpha_2)}\left[\frac{1}{h_2 + (L_2 - L_1)\tan(\alpha_2)} - \frac{1}{h_2}\right]\right\} \quad (12)$$

The total downforce is the sum of (8) y (11).
We can create a spreadsheet in order to calculate this group of equations.

Another example:

$$v_3 = \frac{v_1 \cdot h_1}{h_3}$$

$$v_{x2} = \frac{v_1 \cdot h_1}{h_{x2}}$$

$$\tan(\alpha_2) = \frac{h_3 - h_2}{L_2}$$

$$h_{x2} = x_2 \cdot \tan(\alpha_2) + h_2$$

$$P(x_2) = \frac{1}{2} \cdot \rho \left[\left(\frac{v_1 \cdot h_1}{h_3} \right)^2 - \left(\frac{v_1 \cdot h_1}{x_2 \cdot \tan(\alpha_2) + h_2} \right)^2 \right]$$

$$P_2 = \frac{1}{2} \cdot \rho \left[\left(\frac{v_1 \cdot h_1}{h_3} \right)^2 - \left(\frac{v_1 \cdot h_1}{h_2} \right)^2 \right]$$

$$v_{x1} = \frac{v_1 \cdot h_1}{h_{x1}}$$

$$\tan(\alpha_1) = \frac{h_2 - h_1}{L_1}$$

$$h_{x1} = x_1 \cdot \tan(\alpha_1) + h_1$$

$$P(x_1) = \frac{1}{2} \cdot \rho \left[\left(\frac{v_1 \cdot h_1}{h_3} \right)^2 - \left(\frac{v_1 \cdot h_1}{x_1 \cdot \tan(\alpha_1) + h_1} \right)^2 \right]$$

$$F = b \cdot \frac{1}{2} \cdot \rho \left[\left(\int_0^{L1} \left(\frac{v_1 \cdot h_1}{h_3} \right)^2 - \left(\frac{v_1 \cdot h_1}{x_1 \cdot \tan(\alpha_1) + h_1} \right)^2 \right) dx_1 \right.$$

$$\left. - \left(\int_0^{L2} \left(\frac{v_1 \cdot h_1}{h_3} \right)^2 - \left(\frac{v_1 \cdot h_1}{x_2 \cdot \tan(\alpha_2) + h_2} \right)^2 \right) dx_2 \right]$$

$$B = \frac{v_1 \cdot h_1}{h_3}$$

$$C = v_1 \cdot h_1$$

$$T = \tan(\alpha_1)$$

$$H = h_1$$

$$U = \tan(\alpha_2)$$

$$I = h_2$$

$$A = \frac{b \cdot \rho}{2}$$

$$F = A \cdot \left[\left(B^2 \cdot x + \frac{C^2}{T \cdot (H + T \cdot x)} \cdot I_0^{L1} \right) + \left(B^2 \cdot x + \frac{C^2}{U \cdot (I + U \cdot x)} \cdot I_0^{L2} \right) \right]$$

A specific example:

Input data	
Fluid	
Density (kg/m^3)	**1.18145**
Geometry	
L1 (m)	**2**
L2 (m)	**3**
h1 (m)	**1.00E-02**
h2 (m)	**2.00E-02**
h3 (m)	**3.00E-02**
tan α1	0.005
tan α2	0.00333333
b (m)	**1.8**
Boundary conditions	

v1 (m/s)	60
P3 (Pa, gauge)	0

Auxiliary coefficients	
B	20
C	0.6
T	0.005
H	0.01
U	0.00333333
I	0.02

Results		
Force1	-2977.254	Newton
Force 2	-637.983	Newton
Total Force	-3615.237	Newton

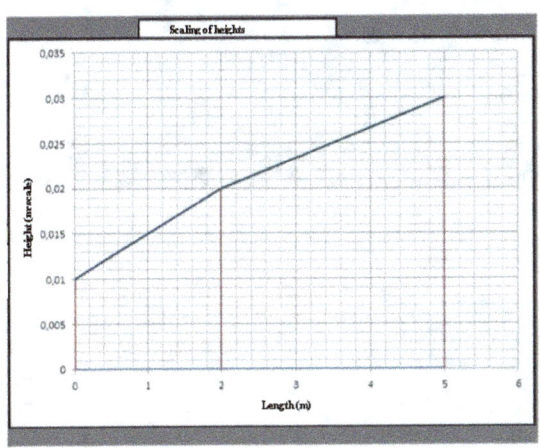

Test 1 and 2 have been performed for different geometry values.

	CFD			EQUATIONS			
	Difusor 1	Difusor 2	Total	Difusor 1	Difusor 2	Total	%Error
Test 1	-2984,58	-641,7	-3626,28	-2984,06	-639,44	-3623,5	0,08%
Test 2	-1299,81	-300,37	-1600,18	-1294,86	-287,75	-1582,62	1,1%

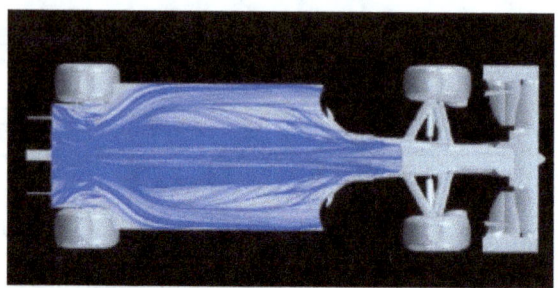

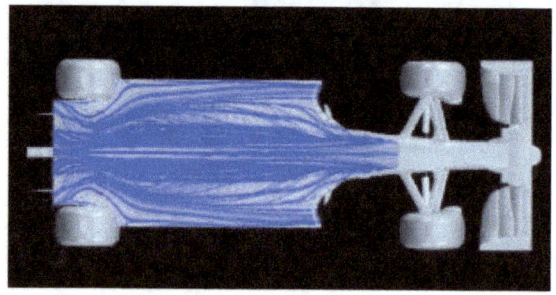

Up:

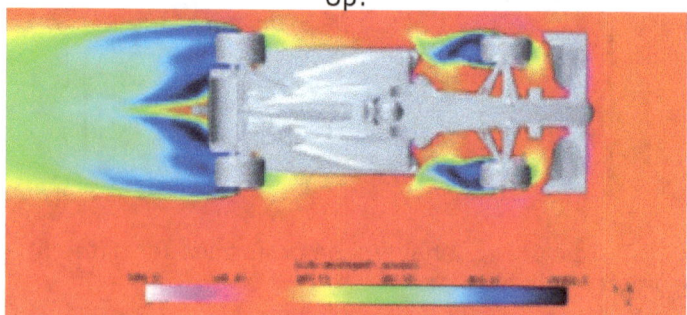

Down:

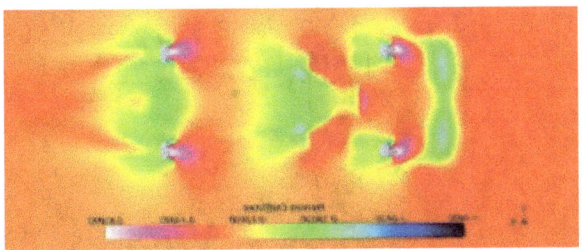

Let's look now at how we can calculate the amount of air mass that goes under the car's floor. This is something necessary to know the effect that it has on the diffuser (with viscosity!):

The parameters that define the problem will be the following:

h (height from the floor) = 0.1m
b (width of the floor) = 2m
V_∞ (intake velocity) = 40m/s
P (density) = 1.225 kg/m^3
μ (viscosity) = 1.8·10^{-5} kg/m·s

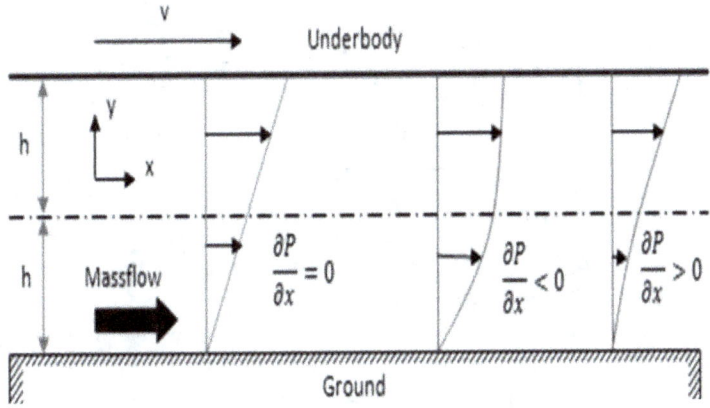

The velocity profile's function is determined by the following expression:

$$u(y) = \frac{1}{2} \cdot \left[\frac{1}{\mu}\frac{dP}{dx}\right] \cdot y^2 + \frac{V_\infty}{2 \cdot h} \cdot y + \frac{\mu \cdot V_\infty - \frac{dP}{dx} \cdot h^2}{2 \cdot \mu}$$

Considering that $\frac{dP}{dx} = 0$ and replacing values, we arrive to the following expression:

$$u(-0.05) = \frac{40}{2 \cdot 0.05} \cdot (-0.05) + \frac{1.8 \cdot 10^3 \cdot 40}{2 \cdot 1.8 \cdot 10^3} = 0 \ m/s$$

$$u(0.05) = \frac{40}{2 \cdot 0.05} \cdot (0.05) + \frac{1.8 \cdot 10^3 \cdot 40}{2 \cdot 1.8 \cdot 10^3} = 40 \ m/s$$

Besides, the mass flow through the section which is formed by the space between the track and car's floor:

$$\iint \rho \cdot \bar{v} \cdot \bar{n} \cdot dA = 0$$

Where " $\bar{n}$ " is a vector perpendicular to the surface differential dA.

Where

$$\bar{v} = u(y) \cdot \hat{\imath}$$

We integrate:

$$\int_{-h}^{h} \rho \cdot \left[\frac{1}{2} \cdot \left[\frac{1}{\mu}\frac{dP}{dx}\right] \cdot y^2 + \frac{V_\infty}{2 \cdot h} \cdot y + \frac{\mu \cdot V_\infty - \frac{dP}{dx} \cdot h^2}{2 \cdot \mu}\right] \cdot dy$$

$$= \frac{h \cdot \rho \cdot \left[3 \cdot \mu \cdot V_\infty - 2 \cdot h^2 \cdot \frac{dP}{dx}\right]}{3 \cdot \mu}$$

As $\frac{dP}{dx} = 0$. Considering the width of the floor, the integral remains as follows:

$$\dot{m} = b \cdot h \cdot \rho \cdot V_\infty = (2 \ m) \cdot (0.05 \ m) \cdot (1.225 \ kg/m^3) \cdot (40 \ m/s) = 4.9 \ \frac{kg}{s}$$

Obviously, the head losses should be noticed. The use of these head losses leads us to define the generalized Bernoulli's principle. These losses reflect the loss of flow energy when it travels through different elements. As the duct becomes longer, the pressure decreases (P_1, P_2 y P_3):

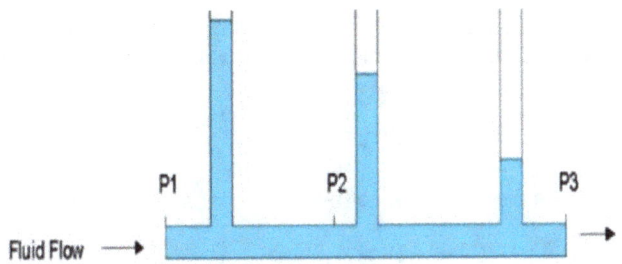

The head losses are due to the fluid friction with the walls of the "duct" through which it circulates and they increase as its velocity increases. As the continuity equation for incompressible fluids has to be true:

$$v_1 \cdot A_1 = v_2 \cdot A_2$$

In this case, $A_1 = A_2$. This expression is true: $v_1 = v_2$. Therefore, the head losses are reflected in the pressure losses, because:

$$v_1{}^2 - v_2{}^2 = 0$$

The head losses between 1 and 2, when they have the same section, are determined by:

$$\Delta Q_{head\ losses} = P_1 - P_2$$

$$\Delta Q_{head\ losses} = \rho \cdot \frac{v_1{}^2}{2} \cdot \left(1 - \frac{A_1{}^2}{A_2{}^2}\right) + P_1 - P_2$$

Let's apply the Bernoulli's principle on a car:

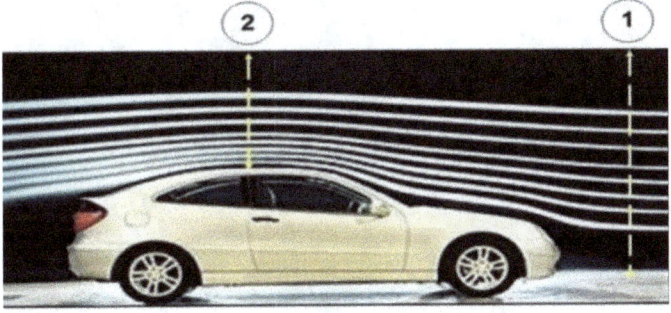

Let's look at, "A" is the area, "P" is the pressure, and "v" is the velocity of the fluid:

Flow: $v_1 \cdot A_1 = v_2 \cdot A_2 \rightarrow \quad v_2 = \dfrac{A_1}{A_2} \cdot v_1$

Therefore, the following expression is true:

$$A_1 > A_2 \rightarrow v_2 > v_1$$

If we operate, we arrive to the following:

$$P_2 = P_1 - \rho \cdot \dfrac{v_2{}^2 - v_1{}^2}{2} \quad \rightarrow \quad P_2 < P_1$$

Let's look at the static equation. We start again from the Euler equation (Navier-Stokes for inviscid fluids):

$$\rho \cdot \left[\dfrac{\partial \bar{v}}{\partial t} + (\bar{v} \cdot \nabla) \cdot \bar{v} \right] = -\nabla P - \rho \cdot \nabla \emptyset_{grav}$$

As this is a static case, there is not movement:
$$\bar{v} = 0$$
We arrive to the following:

$$-\nabla P - \rho \cdot \nabla \emptyset_{grav} = 0$$

With this, we arrive to the hydrostatic equilibrium equation:

$$\nabla P = -\rho \cdot \bar{g}$$

We could describe it as follows:

$$dP = \rho \cdot g \cdot dz$$

We integrate:
$$P = \rho \cdot g \cdot h$$
Where h is the height.

Now, let's look at how we can deduce the same equation or principle form the Navier-Stokes equations. We will work with a vertical coordinate because we would do the same with the rest:

$$\rho \cdot \left(\frac{\partial v_z}{\partial t} + v_x \cdot \frac{\partial v_z}{\partial x} + v_y \cdot \frac{\partial v_z}{\partial y} + v_z \cdot \frac{\partial v_z}{\partial z}\right) = -\frac{\partial P}{\partial z} + \mu \cdot \left(\frac{\partial^2 v_z}{\partial x^2} + \frac{\partial^2 v_z}{\partial y^2} + \frac{\partial^2 v_z}{\partial z^2}\right) + \rho \cdot g_z$$

In the case that we are studying (the body is in a static state), this equation reduces to:

$$\frac{\partial P}{\partial z} = \rho \cdot g_z \; ; \; z \rightarrow h$$

If we integrate, we obtain:
$$P = \rho \cdot g \cdot h$$

As we can see, we have arrived to the same relation with different reasoning.

Let's look at another reasoning to calculate the same relation.

Let's suppose we have a surface with an area "A" which is submerged in a fluid at a "h" depth. "P" is the weight of the "A" section column of fluid that the surface supports:

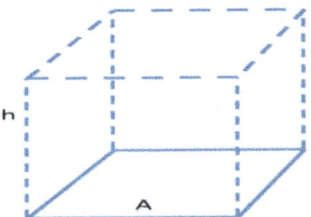

The pressure on the "A" surface will be determined by the force which is applied to this surface, in this case, the weight of the water column with "h" height, per unit of area. In other words,

$$P = \frac{m \cdot g}{A}$$

We can write the density of this fluid in the following way:

$$\rho = \frac{m}{A \cdot h}$$

If we operate conveniently, we obtain again the same relation:

$$\rho = \frac{P}{g \cdot h} \rightarrow P = \rho \cdot g \cdot h$$

Sample:

We found that the speed of water in a hose increased from 1.96 m/s to 25.5 m/s going from the hose to the nozzle. Calculate the pressure in the hose, given that the absolute pressure in the nozzle is 1.01×105 N/m2 (atmospheric, as it must be) and assuming level, frictionless flow.

Strategy:

Level flow means constant depth, so Bernoulli's principle applies. We use the subscript 1 for values in the hose and 2 for those in the nozzle. We are thus asked to find P1.

Solution:

Solving Bernoulli's principle for P1 yields:

$$P_1 = P_2 + \frac{1}{2}\rho v_2{}^2 - \frac{1}{2}\rho v_1{}^2 = P_2 + \frac{1}{2}\rho(v_2{}^2 - v_1{}^2)$$

Substituting known values:

Discussion:

$$P_1 = 1.01 \cdot 10^5 \frac{N}{m^2} + \frac{1}{2}\left(10^3 \frac{kg}{m^3}\right)\left[\left(25.5\frac{m}{s}\right)^2 - \left(1.96\frac{m}{s}\right)^2\right] = 4.24 \cdot 10^5 \; N/m^2$$

This absolute pressure in the hose is greater than in the nozzle, as expected since *v* is greater in the nozzle.

The pressure *P2* in the nozzle must be atmospheric since it emerges into the atmosphere without other changes in conditions.

Another very important sample:

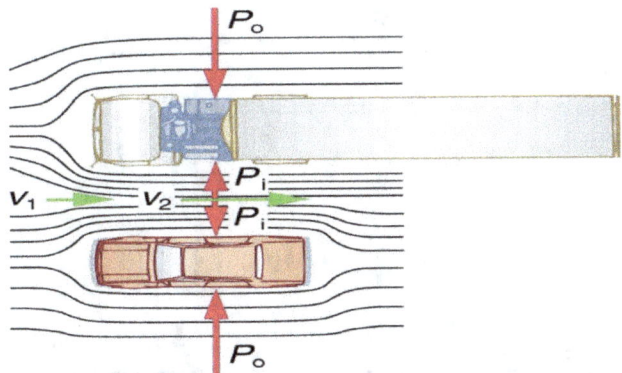

If the difference between P_o and P_i is huge, it can be a very dangerous situation!

VENTURI: PITOT TUBE

The Venturi effect is a direct consequence of the Bernoulli's principle. It says that if there is a reduction in the airflow section, there will be a pressure (dynamic) difference. This pressure difference will be proportional to the heights difference of the fluid column, as we can see on the following graphic:

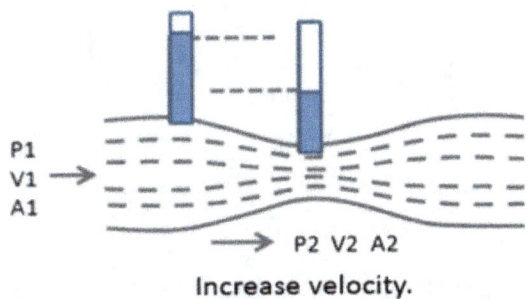

Increase velocity.
Decrease pressure.

In this way, we will be able to know the air velocity, depending on the difference of pressures between both sections (big and small).

The "Pitot" tube is the system that calculates the relative velocity between the car and the airflow as a function of the pressure differences.

Let's look at a sketch of a "Pitot" tube:

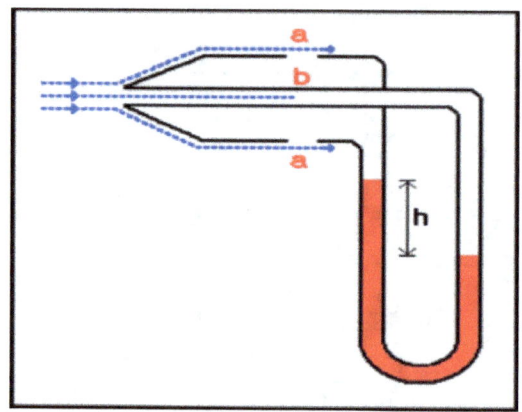

The intakes that are labeled as "a" are called static intakes, and the "b" intake is called dynamic intake (because it is the intake that the airflow changes):

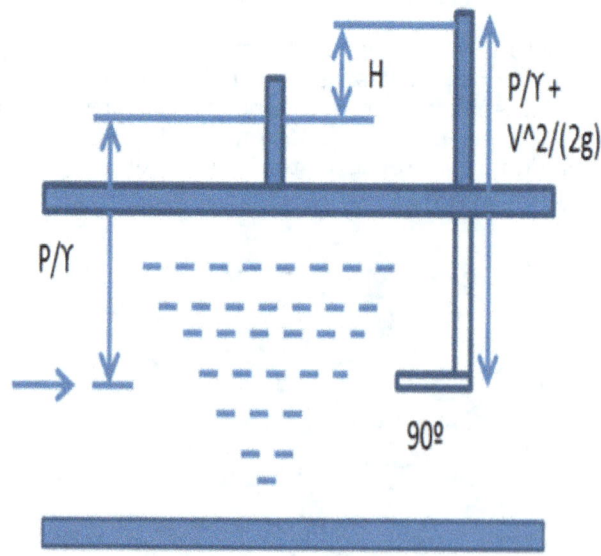

Where:

$$\gamma = \rho \cdot g$$
$$H = \frac{V^2}{2 \cdot g}$$

As we can see, the height difference is proportional to the square of the velocity. Knowing this we can determine the velocity:

$$V = C \cdot \sqrt{2 \cdot g \cdot H}$$

We multiply it by constant "C" to correct the "intake" effect (of shape) of the air through the intake (there is always a head loss and it is necessary to correct it).

The height differences "H" is also important, because it is proportional to the difference in the squares of the velocity.

In fact, if we need (particularly in *tests*) to know perfectly the velocity with respect to the air, we have to place the "Pitot" on a place that it is not affected by turbulences in order to avoid interferences and mistakes in measurement. This is the reason why, in *test* season, we can see that the teams place it on weird places, distant from possible turbulences and interactions with any car element, as we can see on the following picture:

Let's look at an example of measurement of velocity with a "Pitot" tube:

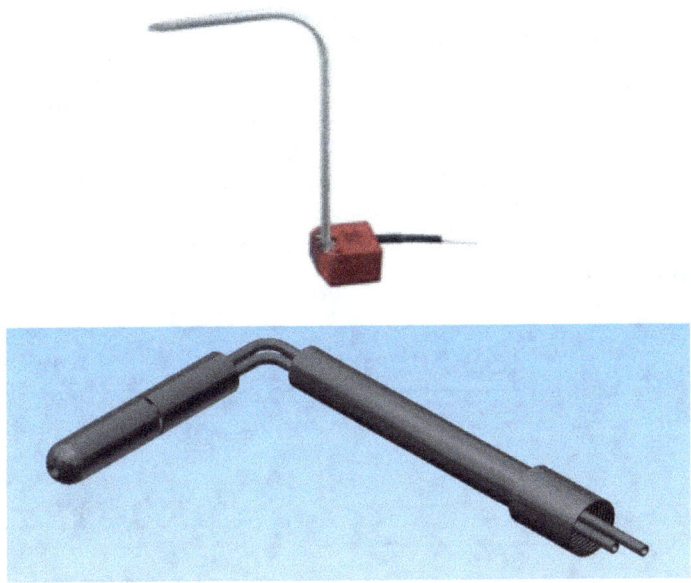

Let's think of a test in the famous "Pikes Peak" circuit, the conditions will be the following:
Toward the top of the mountain, what velocity is required to make the stagnation/static pressure ratio equal to 1.0082 ?

Height = 4300m
P (Density) = 0.79 kg/m³
P_{static} (Atmospheric pressure) = 59 kPa

The pressures relation is determined by the following relation:

$$P_{stagnation} = P_{static} + \frac{1}{2} \cdot \rho \cdot V_2^{\,2}$$

We can also describe it in the following way:

$$\frac{P_{stagnation}}{P_{static}} = 1 + \frac{\frac{1}{2} \cdot \rho \cdot V_2^2}{P_{static}}$$

The velocity will be determined by:

$$V_2 = \sqrt{2 \cdot \frac{P_{static}}{\rho} \cdot \left[\frac{P_{stagnation}}{P_{static}} - 1\right]}$$

Replacing the problem data in the different equations:

$$V_2 = \sqrt{2 \cdot \frac{59000}{0.79} \cdot [1.0082 - 1]} = 35 \, \frac{m}{s}$$

$$P_{Dinamic} = \frac{1}{2} \cdot \rho \cdot V_2^2 = 0.484 \, kPa$$

$$P_{stagnation} = P_{static} + P_{dinamic} = 59.484 \, kPa$$

The pressure decrease that is caused in the section reductions can be exploited in different ways, for example, to suck, through a duct, a fluid (in this case, paint (or chimney as before)):

Another variation geometric about Venturi effect, in order to measure the caudal and speed is:

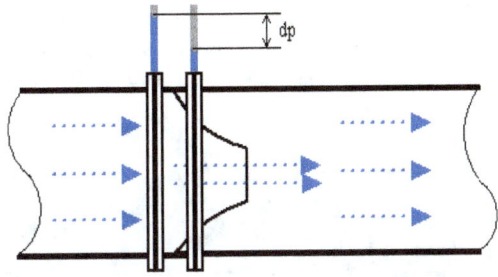

Another application of the Venturi effect is "to help" an airflow move to a different area. To do this, we can suction or inject air (coming from another zone) to the flow. To do this, a series of holes or cracks can be done to connect both zones:

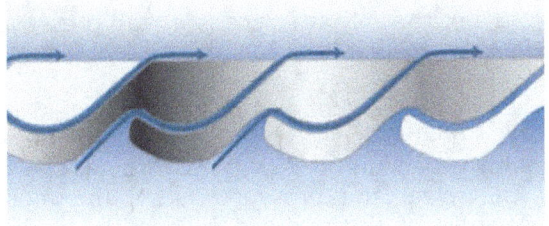

In motorsport, there can be several variations or applications of this phenomenon. Let's imagine a diffuser: inside this diffuser, a relatively low pressure is generated, and we can use it to suction air from another place. In the following image, we can see what we have talked about:

We have, therefore, two cavities with low pressure and a duct in every cavity, suctioning air from "another" place. They also can be used to inject air? Maybe...

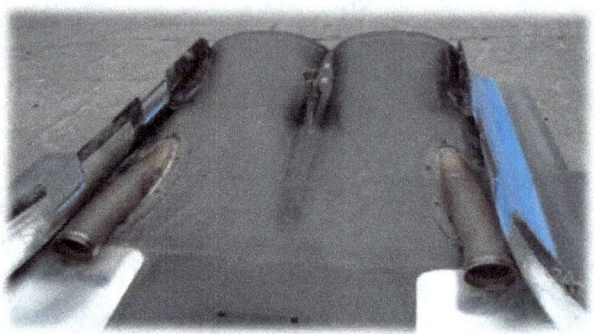

Maybe, the "other" place is a high pressure zone, which we have to eliminate or attenuate... Good idea in this case!

Let's observe that this particular system is called "self-generation system". The higher the velocity with which the airflow goes through the diffuser (the lower the pressure in it), the higher the quantity of air the ducts will suction.

What is more, we can inject air to another zone instead of suctioning air from another zone. It's practically the same, maybe an interpretation of the principle from a different point of view. For example, let's think of Red Bull car and its rear part as an example:

It is the bottom part of the sidepods. The holes that can be appreciated, connect this zone with the upper zone of the end of the diffuser (we will see it in other chapters). This last part is a low pressure zone. Therefore, this system is again a self-generation system. A really good idea!!!

But these inlets, can be also (seem double difusser….):

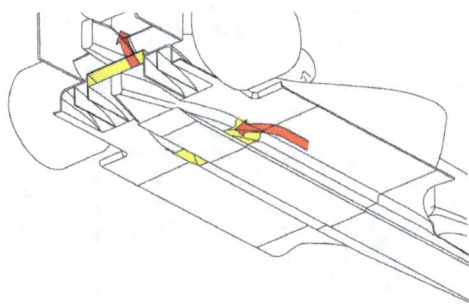

Also, is possible to add air from exhaust, to the same zone (up difusser); this procedure can produce also one skirt (chapter Ground and Difusser):

Another's applications (buildings, etc....):

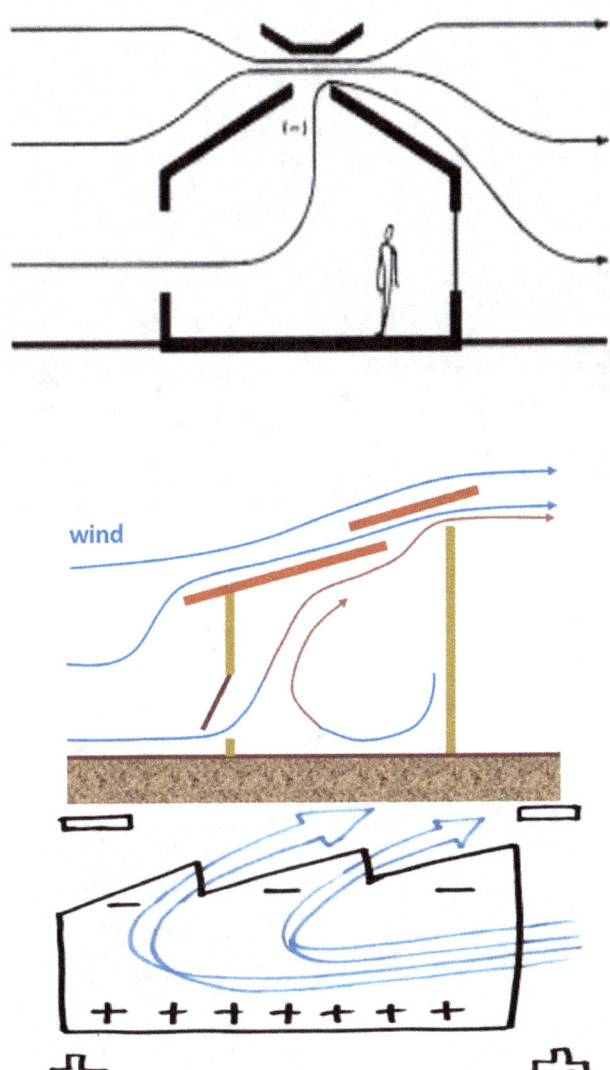

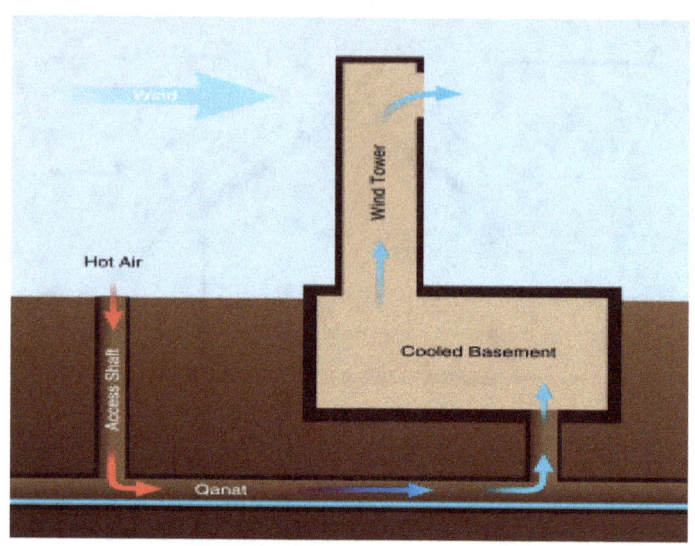

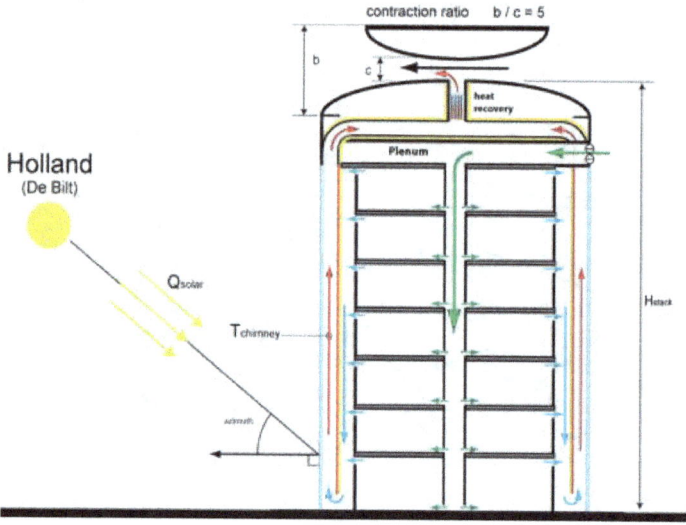

Prevailing Wind

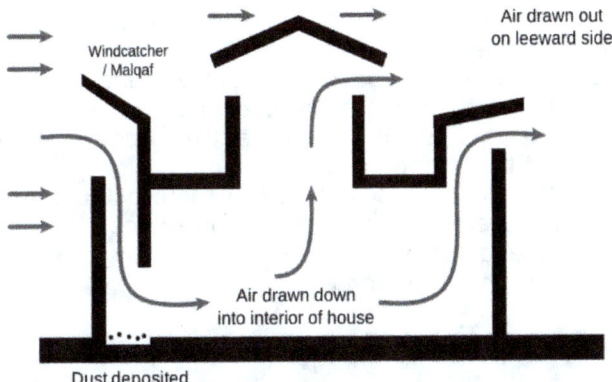

MAGNUS EFFECT

The wheels of a car in motion rotate and because of it, they cause some movements, which can be useful to the car dynamics. We have to consider that the wheels are a big part of the front of the car. Therefore, they cause a lot of drag. A reduction of drag generated by the wheels will cause a total reduction which will be very considerable and important.

Let's think of a ball without rotation: the streamlines which travel around this ball can be represented in the following way:

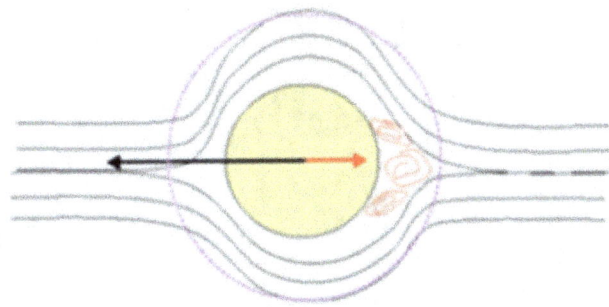

We can observe that there is no difference between the lines at the top and bottom part. Meaning: there is no difference of pressure. Wouldn't this also depend on the Reynolds number? For some transition Reynolds, we would have temporary behaviors, non-symmetric solutions which are known as the von Karman Street, but this is not the case that concerns us to understand the phenomenon.

Now, let's suppose that the same ball is rotating. There is, therefore, a perceptible difference of pressure on the top and bottom of the ball. The air goes from the left to the right in the image:

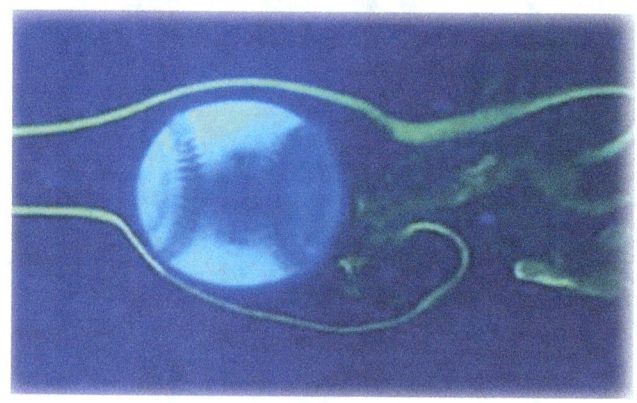

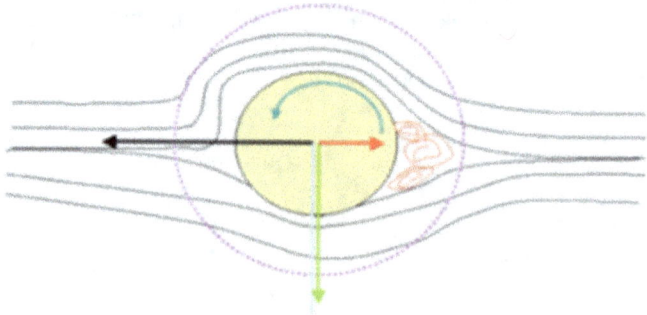

If we turn the ball in the direction indicated with the green arrow and if there is this difference of pressure, the ball will tend to go down: there is lower pressure below because the flow circulates quicker.

Depending on the direction in which the ball rotates, it will go up or down. In other words, the resultant force will have a determined direction, depending on the direction of the ball.

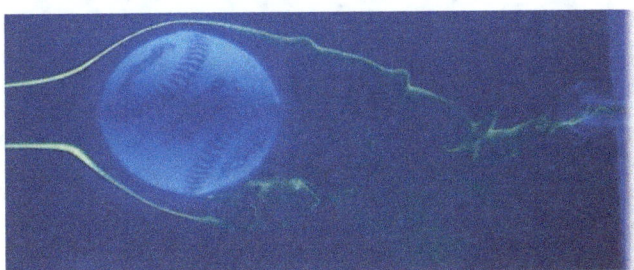

Ideally, we can calculate the push force that is caused in the rotation by the Kutta-Joukowski theorem; in its principle obtained to find the push force which is caused by an infinite cylinder. Let's think of a sphere of a "b" radius which is rotating with an "s" angular velocity on which a flow with "ρ" density and "v" velocity has an impact:

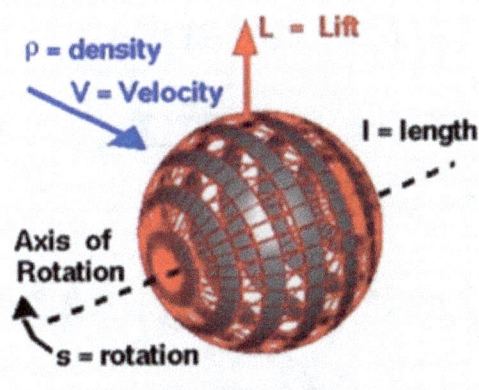

ρ = density
V = Velocity
L = Lift
l = length
Axis of Rotation
s = rotation
b = radius of ball

Lift is defined for a ring of sphere in the following way:

$$L = \rho \cdot \Gamma \cdot V$$

Where "Γ", is the circulation, "V" is the fluid velocity and "ρ", is the density of the fluid:

$$\Gamma(r) = 2 \cdot \pi \cdot r \cdot V(r)$$

$$V(r) = 2 \cdot \pi \cdot r \cdot s$$

$$\Gamma(r) = (2 \cdot \pi \cdot r)^2 \cdot s$$

$$L = (2 \cdot \pi \cdot r)^2 \cdot \rho \cdot V$$

We integrate along the rotation axis to obtain the *lift* of the whole sphere:

$$L = \int_{-b}^{b} (2 \cdot \pi \cdot r)^2 \cdot \rho \cdot r \cdot V \cdot dl$$

We change the variable:

$$r = b \cdot \sin \phi$$

$$l = b \cdot \cos \phi$$

$$dl = -b \cdot \sin \phi \cdot d\phi$$

Replacing:

$$L = (2 \cdot \pi)^2 \cdot s \cdot \rho \cdot V \cdot \int_{0}^{\pi} \sin^3 \phi \cdot d\phi$$

This gives us the resultant lift generated by the sphere:

$$L = \frac{4}{3} \cdot (2 \cdot \pi)^2 \cdot s \cdot \rho \cdot V \cdot b^3$$

Let's get back to the pressures field on the sphere. There is a pressure differences due to:

- The rotation.
- The air viscosity.
- The friction.

Summarizing some ideas:

- The higher the viscosity, the greater the Magnus effect (it will be a higher pressure difference).
- The higher the friction of the ball with the air, the greater the Magnus effect. Therefore, it is possible to change the material or the roughness of the ball to generate more or less "effect" or curvature in its trajectory.
- The higher the turning speed, the higher the capacity to change the direction.

As we can guess, goals on the moon wouldn't be as amazing as on the earth, because on the moon there is no air and the ball cannot be kicked with a curved trajectory...

There are some applications to this phenomenon, for example, with boats. It is possible to cause "artificial candles" (rotating cylinders) to produce feed force.

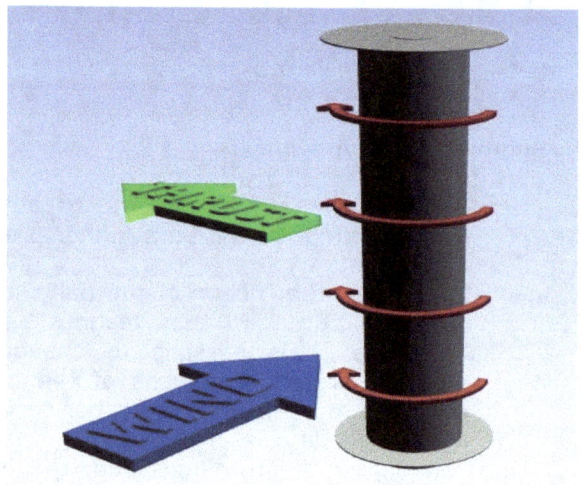

The Magnus effect can only be produced if the wheel is rotating. The golf ball, with its characteristic dimples, reaches a longer distance than the smooth ball.

Each dimple produces a small depression that suctions the flow, delaying the whole boundary layer detachment. If it is rotating, it is possible to delay the detachment even more, with which the reached distance is higher:

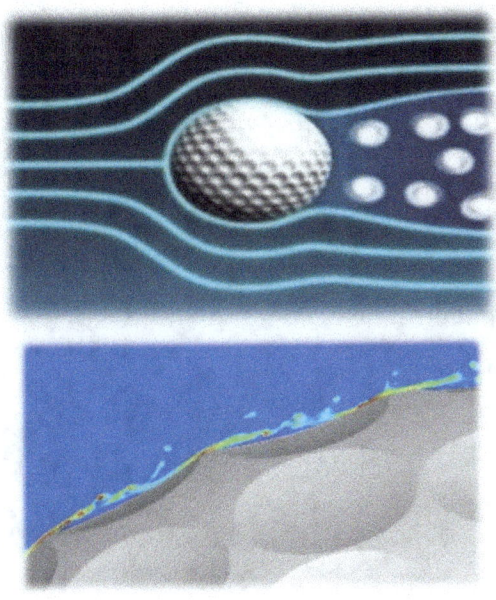

In the following case, although the helmet is not in rotation, the drag decreases. This is due to the holes placed in the front part of the helmet that make the flow become turbulent. This causes, as we will see, a detachment delay. Maybe, it can also reduce the *drag* on the frontal part:

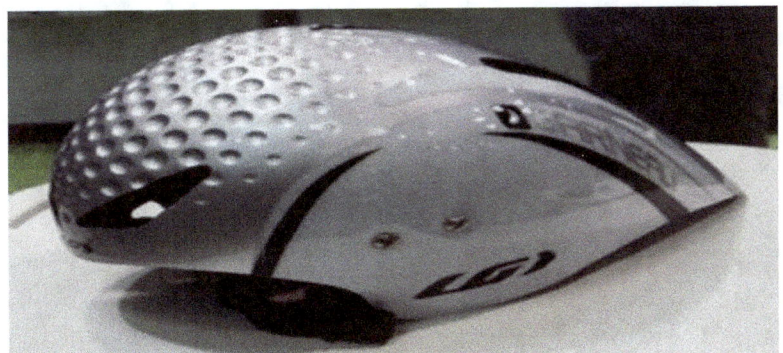

Can you install dimples on cars or other vehicles, to decrease the resistance? False: It is most likely that there is no effect, but only the increase in *drag*.

In a famous TV show, they made dimples on the whole surface of a car. Because dimples were placed in a random order, the car consumed more fuel with dimples than without them:

When I think of the Magnus effect, I like to think of the billiard game. If we want to give a good spin to the ball, we have to kick it on its side, but the most important thing in billiard is the velvet mat: the mat and the ball "hold on" to each other and the mat converts the wheel rotation into a "spin", If instead of a mat, we had, for example, a marble platform, we wouldn't be able to give any spin move to any billiard ball. It is the same reason why a ball cannot be kicked with a spin move on the moon.

Let's think of the most important: the car wheels. The following image is very clarifying when we observe the air movement on the wheels. We obtain some ideas that we have to know:

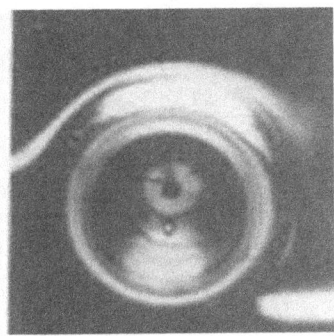

If the wheel doesn't move (image on the left), the airflow goes up. By contrast, if the wheel is in rotation (image on the right), airflow circulates around the wheel, descending downstream afterwards. What does this mean?

By the action-reaction law, if the air goes up, the wheel goes down. In other words, if the wheels are in movement, there is an additional *downforce*, generated by the wheels movement. It is also true that on the frontal part of the wheel, there is a "plug" which is caused by the own wheel. At the end (net amount), the wheel produces lift...

Later, we will see the aerodynamic influence that the following elements have on the car:

- The wheels rotation.
- The *yaw* of the car.
- Steering.

If the wheel rotates, the resistance changes and when the flow around it changes, the downstream effects will be different as well.

Let's look at a table. We can observe the *lift* and *drag* coefficients depending on the wheel size and depending on whether it rotates or not:

Thickness/Diameter ratio	CL	CL (NR)	CD	CD (NR)
0.28	0.579	0.593	0.18	
0.5	0.65	0.75	0.4	0.272
0.612	0.56	0.77	0.48	0.95
0.658	0.6		0.32	0.76

(NR) corresponds to the wheels without movement.

In the following chart, we can observe the aerodynamic resistance difference of a wheel (with and without movement) (car with a frontal area of 2 m^2). The most important thing is not only this fact, but the *yaw* effect with this resistance. Hence, knowing the aerodynamic influence of the wheels in a corner is very important. It is a very important graphic.

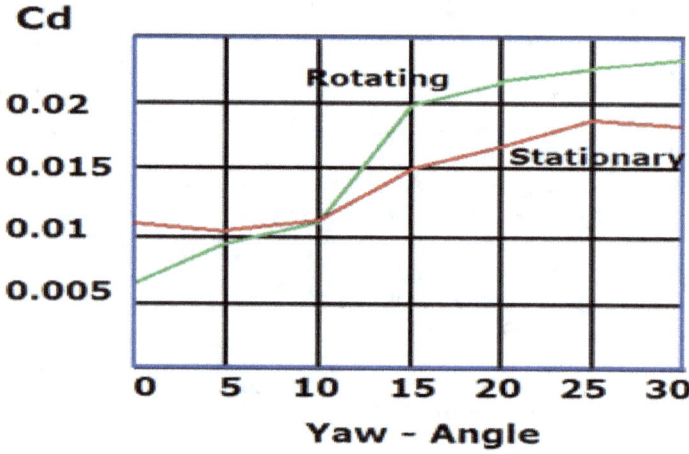

We have already said it but we have to insist on it: The rotation of wheels affects the aerodynamic behavior of the rear part of the car: there can be big variations of behavior on the floor and on the diffuser, for example:

Could there be a system implementing the Magnus effect and the Coanda effect?

Let's suppose that we have a wing and we replace the leading edge with a cylinder in rotation:

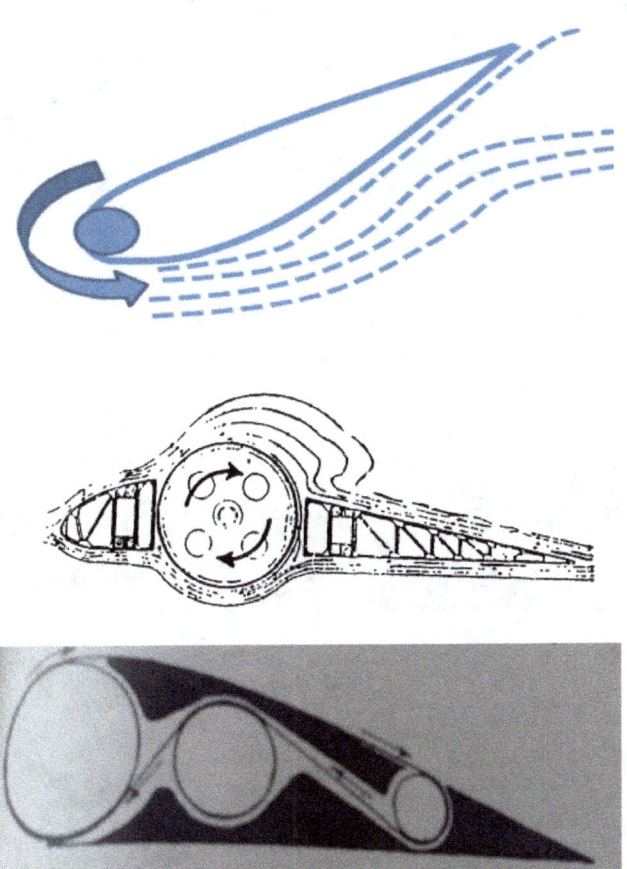

What do we get with this? The flow gains speed "on the lower surface" and travels the whole surface, going out through the trailing edge without detaching from it. In some way, it is similar to the BLS system (injector) that we have already seen. This adhesion is due to the Coanda effect. What is more, there are some planes that use this system to fly:

We have to notice that this plane has flaps on the cylinder of the flow edge. Hence, it has a higher lift by the effect of the blade turbine. In this case, it is a combination between BLS and wing beat:

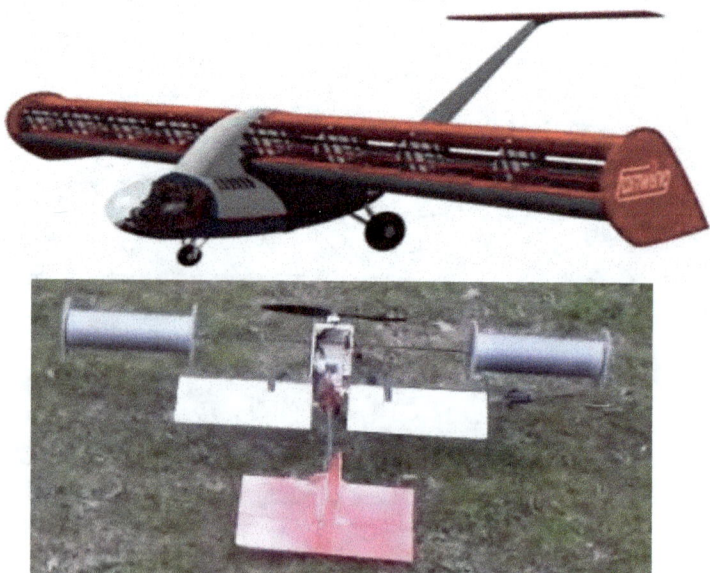

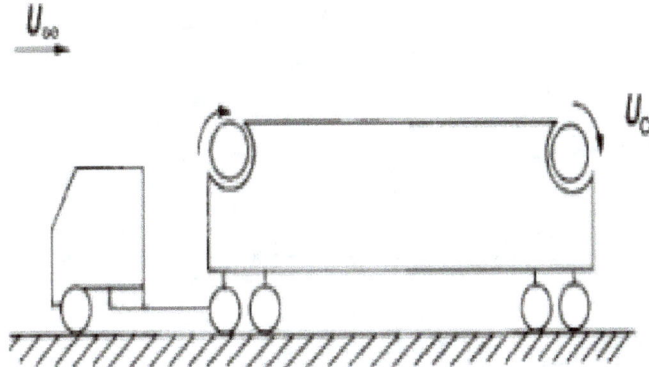

A racing tire has a very particular pressure field. We can represent this pressure field depending on its distance from the floor.

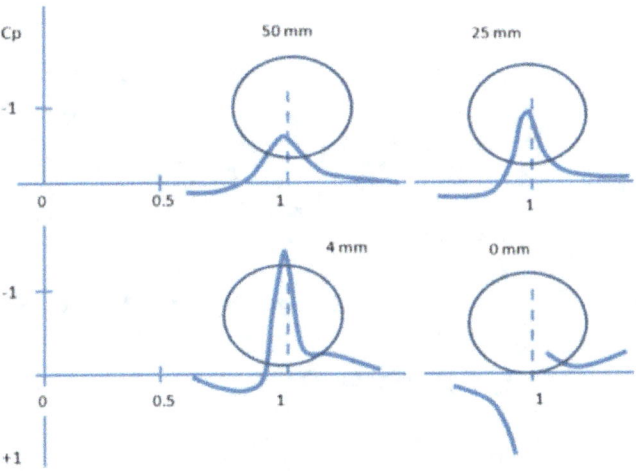

That is:

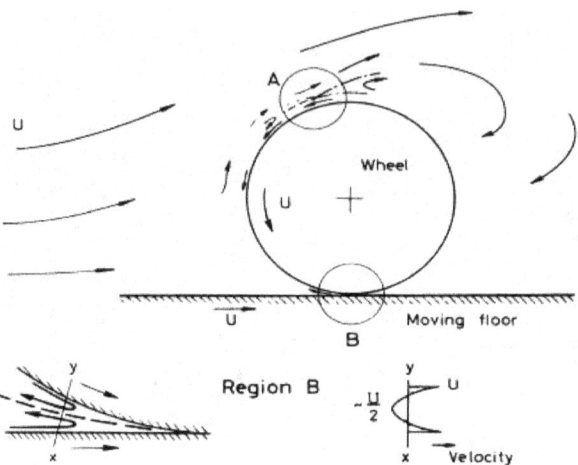

U

A

Wheel

U

Moving floor

U

B

Region B

$-\frac{U}{2}$ U

y

x

y

x Velocity

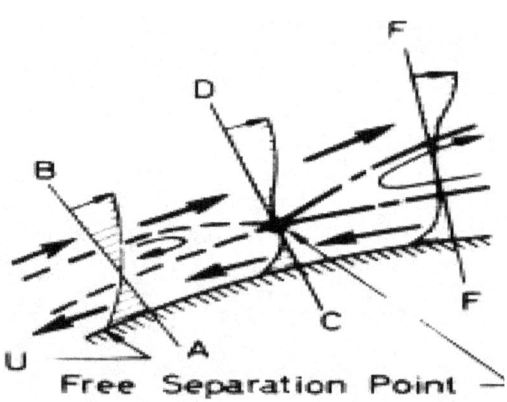

F

D

B

U

A

C

F

Free Separation Point —

Region A

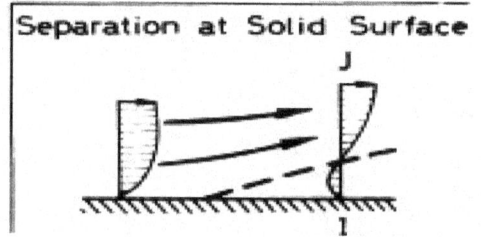

Separation at Solid Surface

J

1

When the tire is in contact with the road, the pressure coefficient is in its maximum. Then, the pressure's net amount is positive (lift). In the other situations *downforce* occurs.

This field of pressure depends on the footprint size or depth. The issue here is how to determine the variation as a function of a size, as for example, the following graphic represents the pressure coefficient in each point of the tire circumference (in degrees), rotating or without movement:

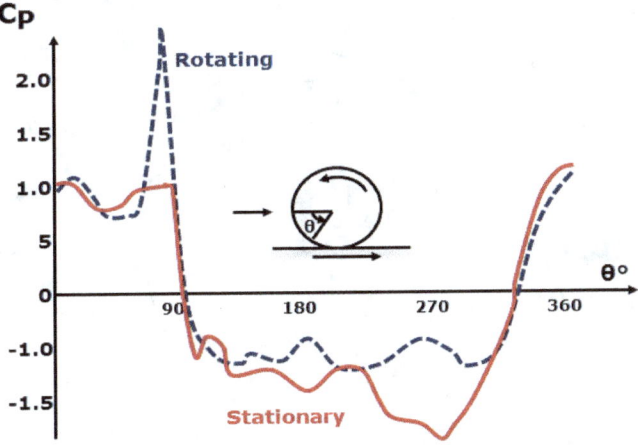

We can obtain the variation of the pressure and the values of *downforce or upforce* and *drag* depending on:

- The angle of attack of a frontal wing in front of the wheel.
- The magnitude of the frontal wing.
- The distance to the floor of the frontal wing.
- The rotation angle or *yaw*.
- "Several" elements which are placed in front of the wheel.
- The roughness of tire.

- Reynolds number

These last 2 factors, are very important. In fact, there is a one effect, named "REVERSE MAGNUS EFFECT".

That is: there is a force in another sense than traditional, against the sense of rotation. This is created by the variation of boundary layer separation, depending some factors.

Depending surfaces slicks or roughen, the force is:

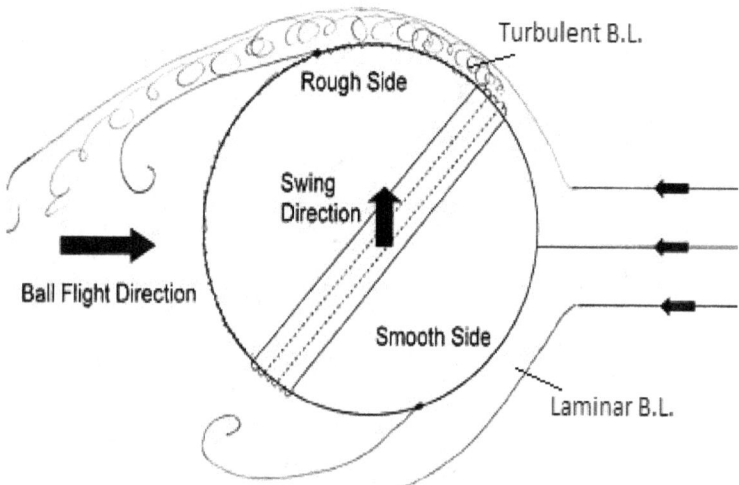

Traditional force.

The separation of boundary layer, is essential in order to create downforce or upforce. That, also, depend of Reynolds number (example about cylinder in rotation):

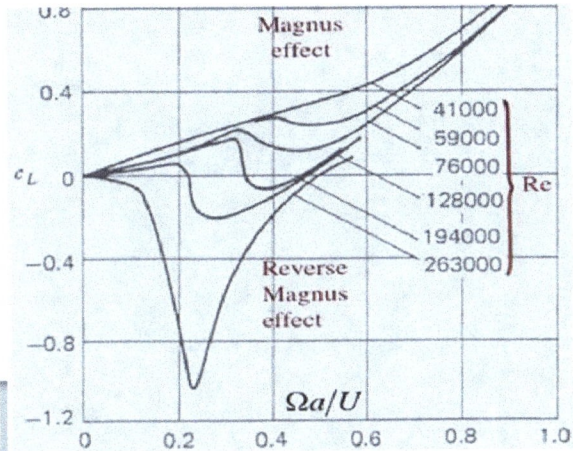

The Magnus force per length unit is:

$$C_L \rho a^2 U \Omega$$

Where "a" is the cylinder radius, "U" its translational speed and "Ω" its rotational speed. So is possible calculate the radius and speed, in order to have downforce or not.

The conclusion is clear:

Depending on these parameters and many more... the pressure on the tire is different. Many studies can been conducted to analyze the pressure variations but the key is that the wheels are in the car and "everything" is an indivisible whole...

In WIFFLE BALL PITCH case, the rotation and the hole in a ball, produce differents ball speed and directions:

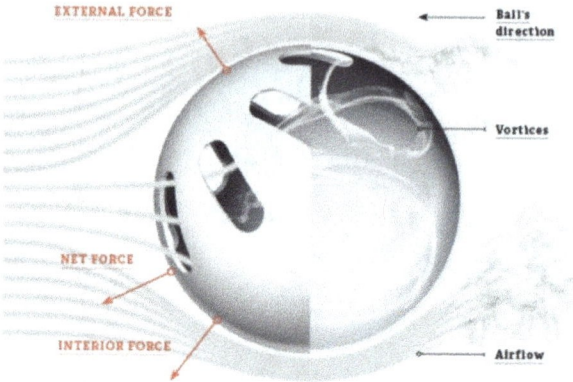

EXTERNAL FORCE

Ball's direction

Vortices

NET FORCE

INTERIOR FORCE

Airflow

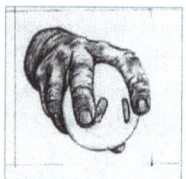

STRAIGHT FASTBALL

CURVE BALL

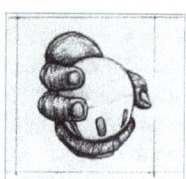

RISER

SCREWBALL

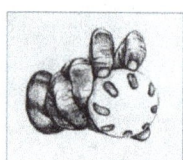

KNUCKLEBALL

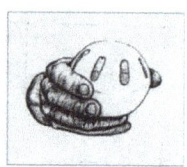

SINKER

6 CURVEBALLS

R/H CURVEBALL
R/H SLIDER
R/H CUTTER
L/H CUTTER
L/H SLIDER
L/H CURVEBALL

CUTTER - SLIDER - CURVE
STANDARD ON ALL SANDLOT SLUGGERS

Another think very important, but very complicate, is the alteration of dynamic vortex or flow in general, depending if wheel is static or in rotation; is complicate but necessary; we see that: the formation and interaction of vortex:

The rotation of wheels, have a lot consequences which is necessary to know; for example, the turbulences, vortex or rare flow behind her.

As a sample, we can see:

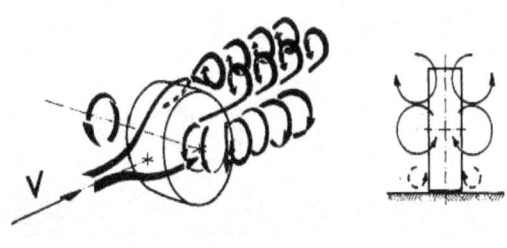

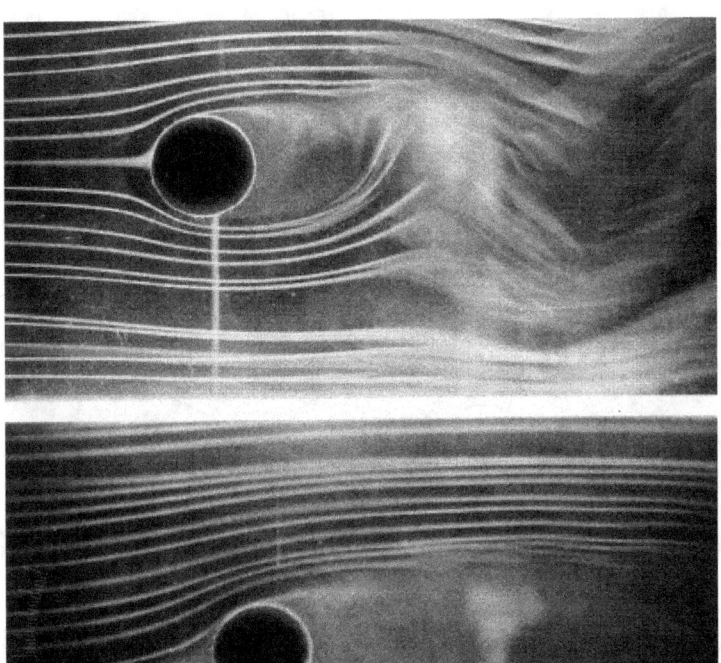

 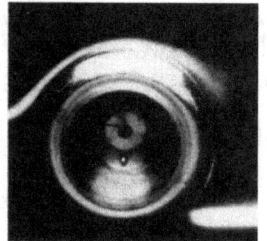

These turbulences in general, produce drag, but also another effects behind (vibrations, interaction between another vortex, help to generate downforce by devices behind, etc....).

The measure of drag and lift, is some very complicate and with a lot variation, depending of lot of thinks (wind tunnel system, etc....):

Width/Diameter	C_D	C_L	Re Number
0.28	0.180 (0.272)	0.579 (0.593)	1.1×10^6
0.50	0.40 (0.95)	0.65 (0.75)	0.2×10^6
0.612	0.48 (0.76)	0.56 (0.77)	5.3×10^5
0.658	0.32	0.60	5.3×10^5

That is the pressure field:

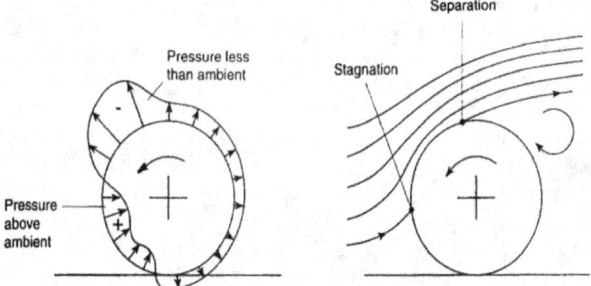

The contact with tire and track, produce vortex too:

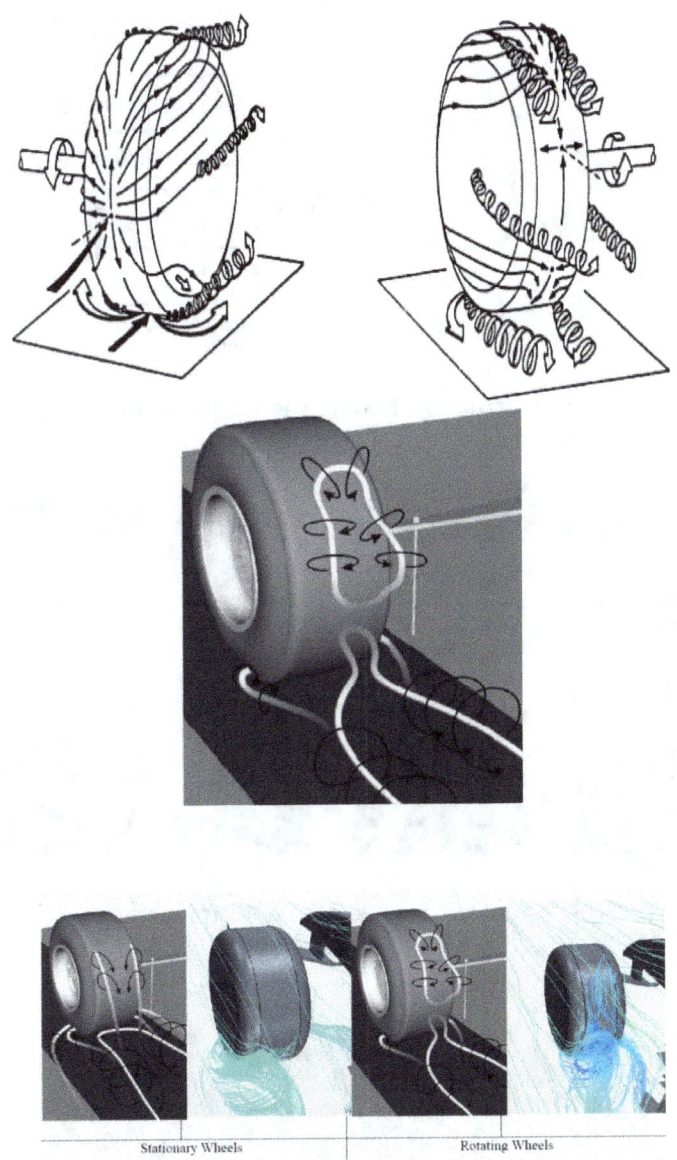

Stationary Wheels   Rotating Wheels

We know already, that one wheel with rotation,
have less lift than without rotation:

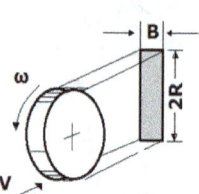

$$c_D = \dfrac{D}{\dfrac{\rho}{2}\,V^2\,B\,2R}$$

	$\omega=0$	$\omega=\dfrac{V}{R}$	
c_D	0.593	0.579	standard rim
c_L	0.272	0.180	
c_D	0.544	0.488	rim faired on both sides, flat cap
c_L	0.296	0.178	

Tire 145 SR 10 Cinturato

Vortex structure behind the front wheel in the wheel arch:

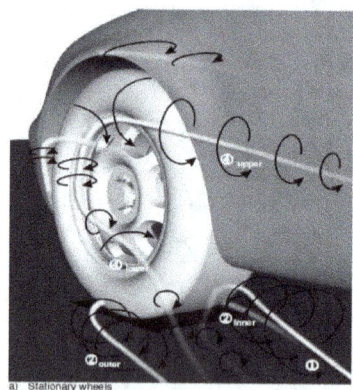
a) Stationary wheels

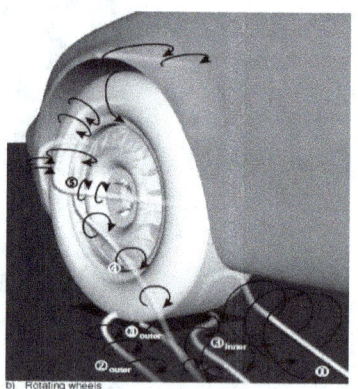
b) Rotating wheels

The wheel rims, have also a lot importance and influence; that is:

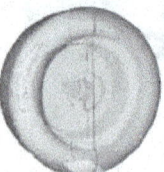

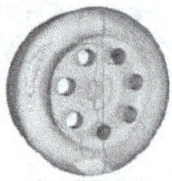

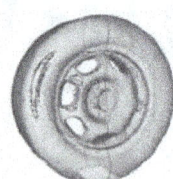

closed rim   simplified rim   detailed rim

Influence of Wheel Rotation on Lift
($\Delta\,C_l = C_l$,rotating $- C_l$, stationary)

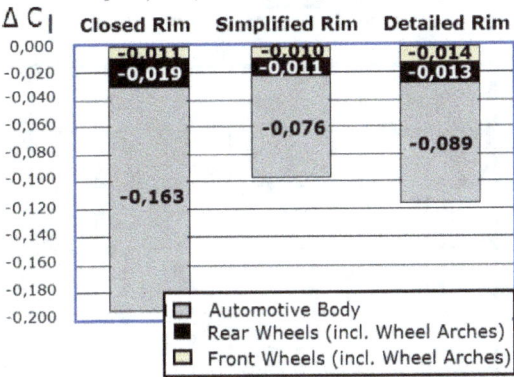

Influence of Wheel Rotation on Drag
($\Delta C_d = C_d$,rotating $- C_d$, stationary)

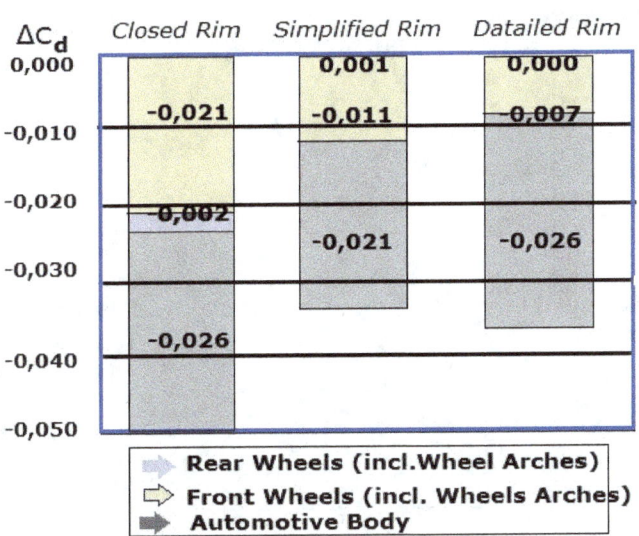

The yaw flow angle in the wheels, is important:

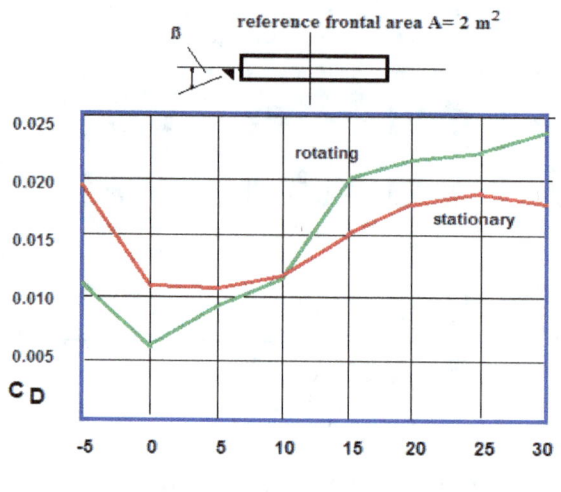

reference frontal area A= 2 m^2

About the vortex created behind, and the speed field created:

Vortex center:

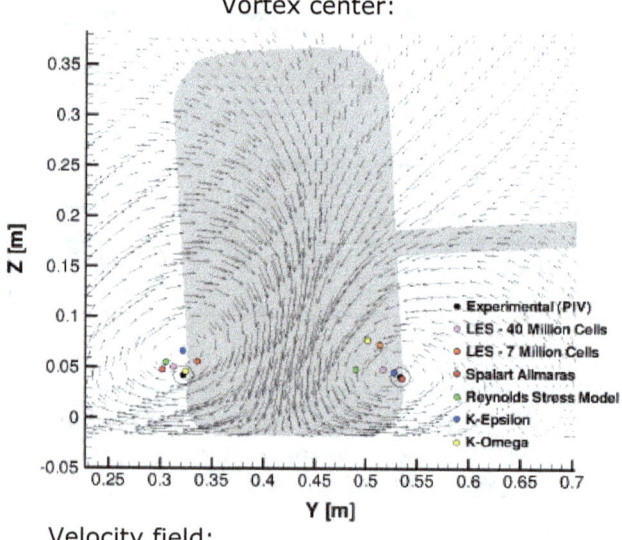

- Experimental (PIV)
- LES - 40 Million Cells
- LES - 7 Million Cells
- Spalart Allmaras
- Reynolds Stress Model
- K-Epsilon
- K-Omega

Velocity field:

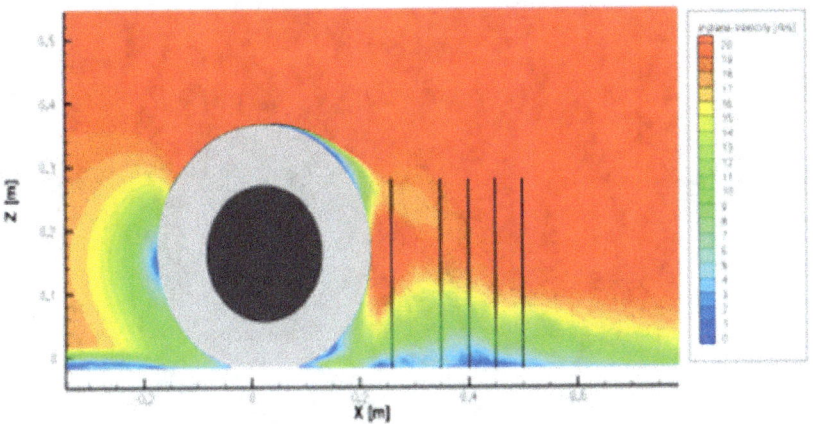

Center-plane velocity comparison (Experiment - Black Dots, LES at 3.25° - Red, RANS at 3.25° with 5% TI - Blue, RANS at 3.25° with 1% TI - Orange, RANS at 2.5° with 5% TI - Green)

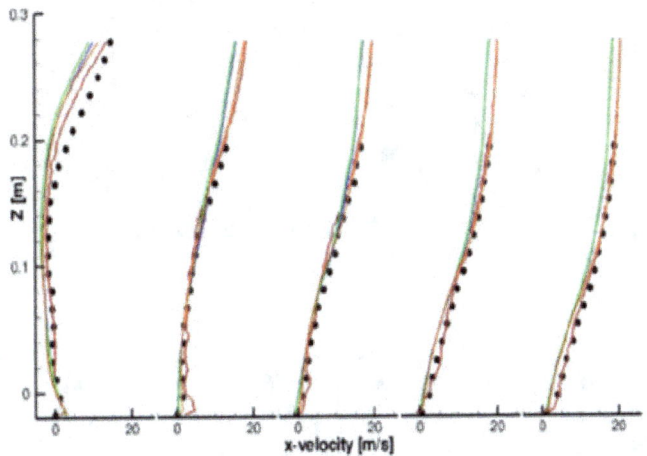

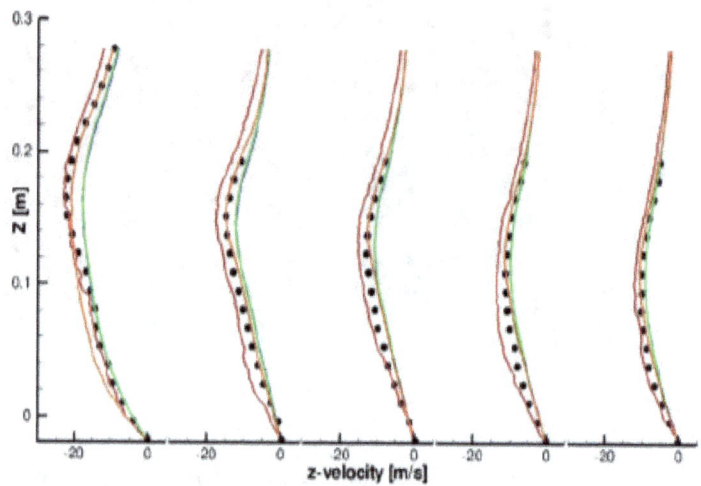

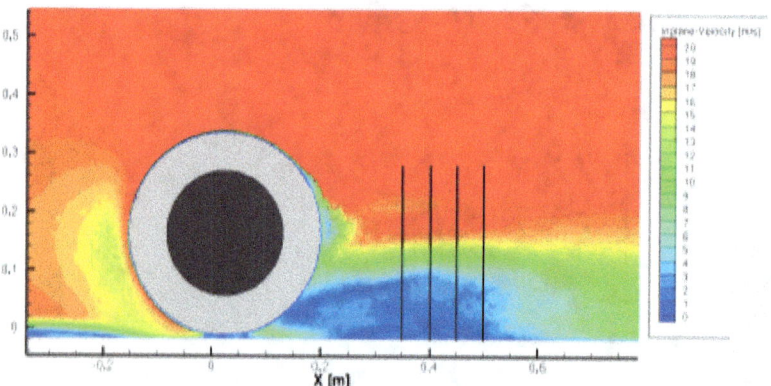

Velocity comparison of a plane y/W=0.5 outboard from the center-plane (Experiment - Black Dots, LES at 3.25° - Red, RANS at 3.25° with 5% TI - Blue, RANS at 3.25° with 1% TI - Orange, RANS at 2.5° with 5% TI - Green)

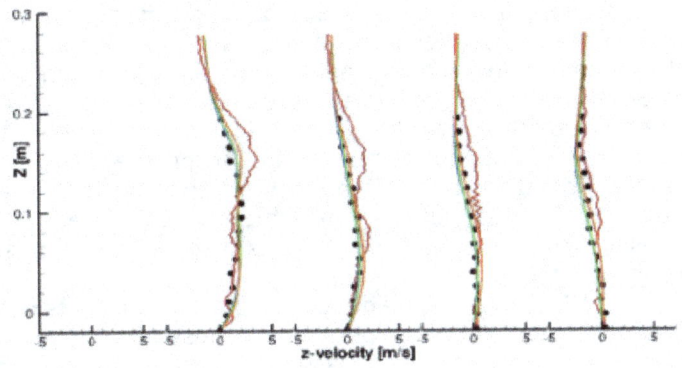

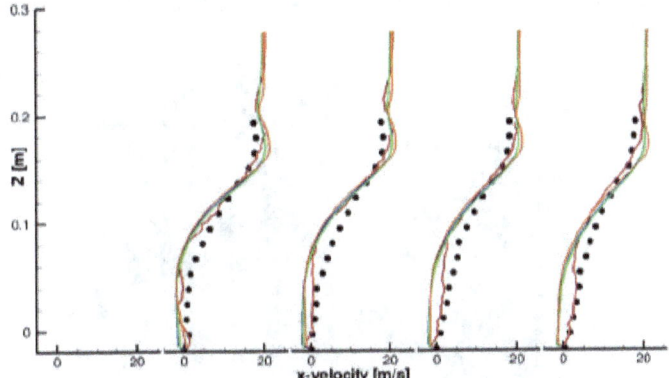

Conclusions about:

Know the flow behind the wheels is very complicate, but necessary; also:

- Lift.
- Drag.
- Influence devices behind.
- Etc....

COANDA EFFECT

We have to study this effect to define all the important effects and properties of the air. It is maybe one of the most important effects but, in reality, all the effects form a group of "pieces" or inseparable properties.

First of all and "almost" as a summary, we have to say that this effect depends on the viscosity and/or the friction. The higher the friction is, the higher the Coanda effect will be. The friction "drags" and "sticks" the fluid to the surface!

The Coanda effect (in honor of the Romanian Henry Coanda), is the tendency of the air to remain adhered to the surface on which it circulates. This attachment will remain till there is a force that opposes to it with enough value.

We have an image explaining this: a water jet, without touching a container, doesn't change its trajectory till it touches "slightly" the container. The water jet tries to cover the recipient following its surface. The Coanda effect is very sensitive to small changes in the initial conditions:

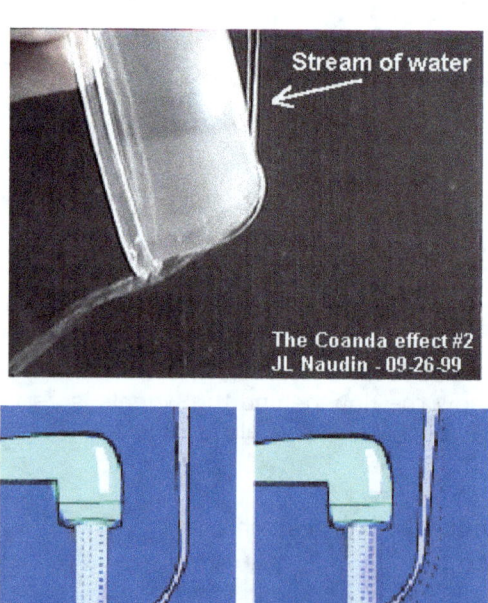

Stream of water

The Coanda effect #2
JL Naudin - 09-26-99

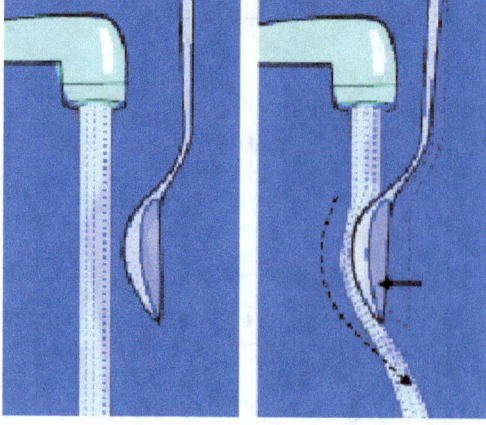

What can we do with this effect?

For example, in order to transfer "further" an airflow from an air conditioning *split*, we can attach it to the wall, "falling" much further than if we tried to inject directly the stream:

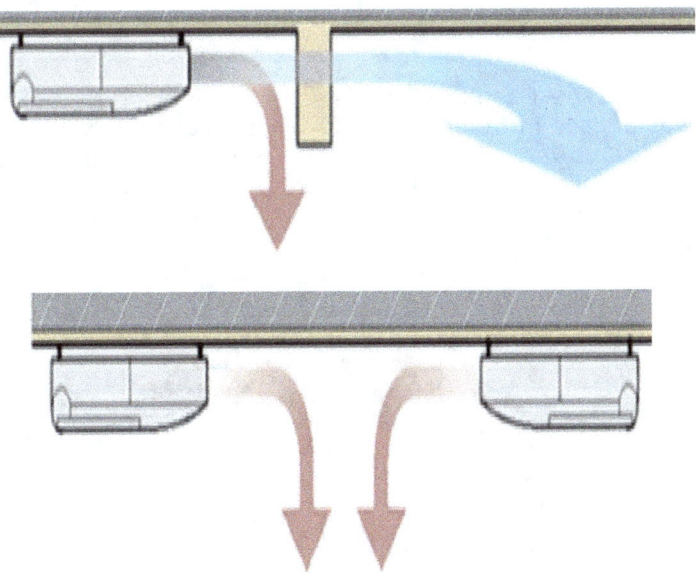

YFCC 'COANDA' CASSETTE UNITS

In order to transport air to all the interior parts of the car circulating adequately:

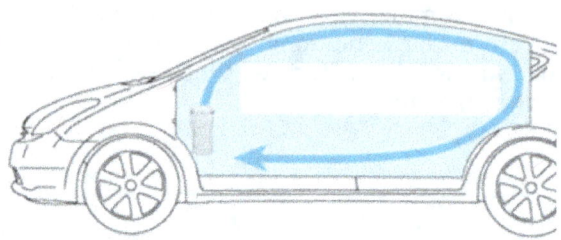

The Coanda effect is known for a long time, It was applied in freight aircrafts. These planes need a high rising power. Their smart technique consisted of transferring the flow of the reactor directly to the wing. The wing redirected the entire stream downward, generating a lot of lift:

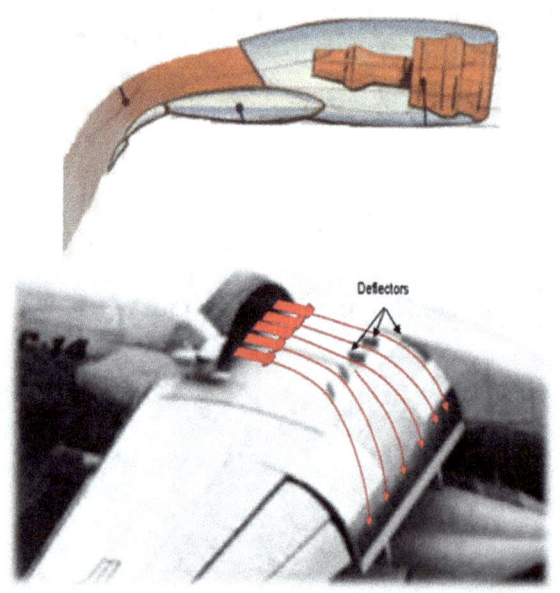

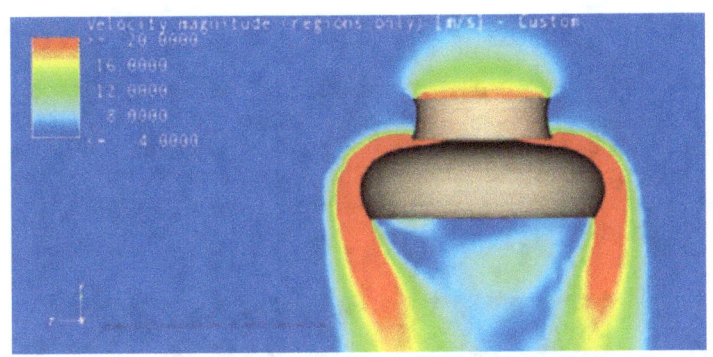

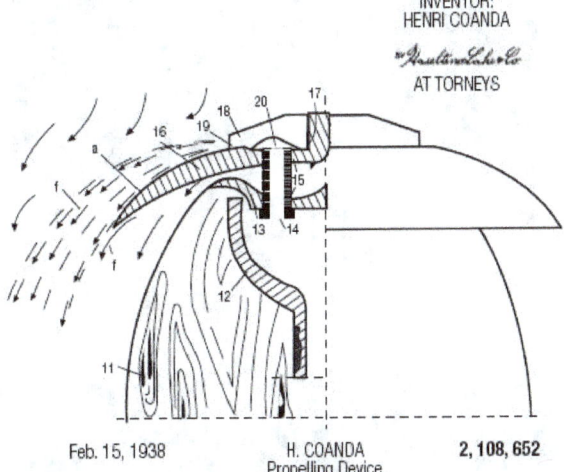

INVENTOR:
HENRI COANDA

AT TORNEYS

Feb. 15, 1938   H. COANDA   2,108,652
                   Propelling Device
                   Filled Jan, 10,1936

This is really important if we want to transport air from one place to another, despite the barriers which block the air. Let's imagine a candle behind a bottle: despite the bottle, if we blow, we can extinguish the candle. The airflow circulates around the bottle, affecting the candle:

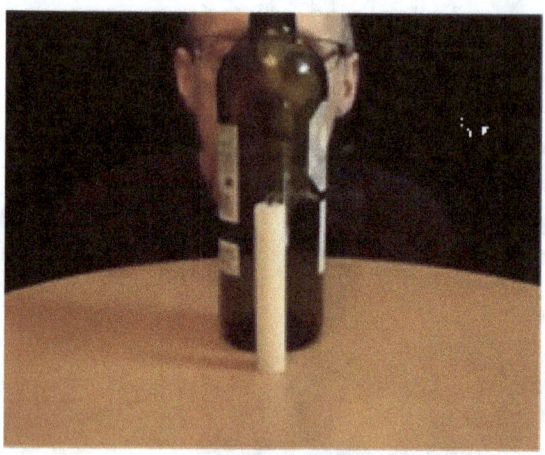

Let's imagine that we have to divert an air flow to an area of the car, usually placed in the backside. When diverting a flow, we can do it in two ways:

1) Placing "solid" deflectors, as plates or something similar, so that the flow when touching them diverts its "normal" trajectory.
2) Using the Coanda effect, making the flow "travel" along a surface and bending it to divert the flow.

The advantages of the second method are many. We have control on the flow anytime. Besides, if we compare this method with the first one, drag is smaller. There is only frictional resistance.

Why is this tendency produced?

When a flow passes on a surface, if this surface bends itself, on the sinus of the flow (which is attached to the surface) a depression develops. This depression maintains the flow attached. As the surface bends itself, this depression starts again; suctioning and maintaining the flow attached. However, everything has its limit...

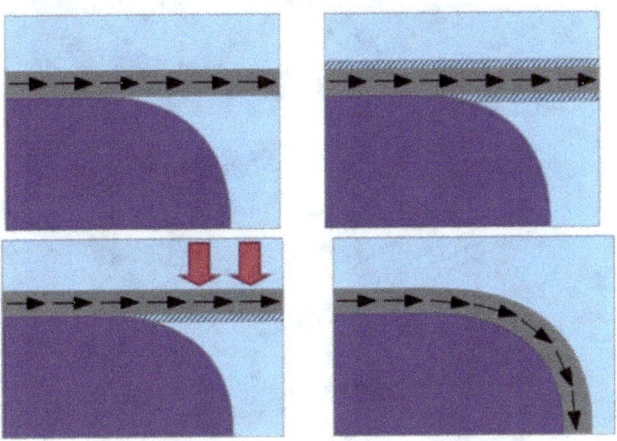

In competition, the Coanda effect is used in several fields and places:

- In order to improve the functioning of the rear wing.
- In order to improve the functioning of the diffuser.

- In order to improve the cooling of rear brakes.
- To canalize correctly the engine exhaust or other air flow.
- In order to decrease the drag.
- Radiators, etc.

We always use surfaces with a goal:

To transport the flow a long distance. We don't use it to channel the flow from the rear-view mirror to the same rear-view mirror (little distances).

We can see the application of this effect in the channeling of the air to the upper rear part of the diffuser: The sidepods are used to achieve this goal:

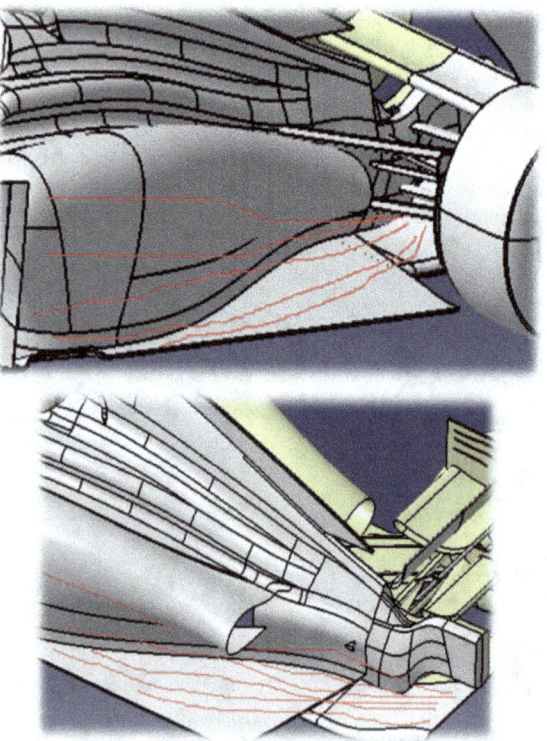

The vertical piece that normally is placed at the beginning of the sidepod, may be is the first to help the flow to adhere to it:

Another device installed by Mc Laren in 2016 season, is to use the air inlet in sidepod, in order to attached it in a sidepod surface (Coanda); that is:

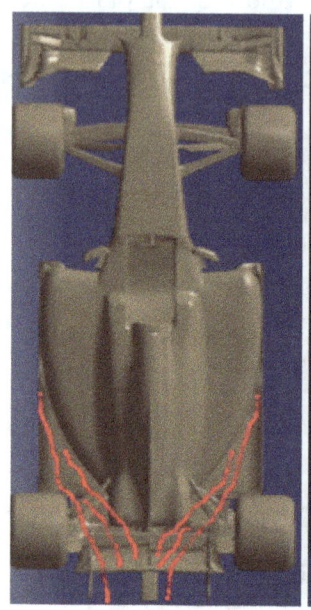

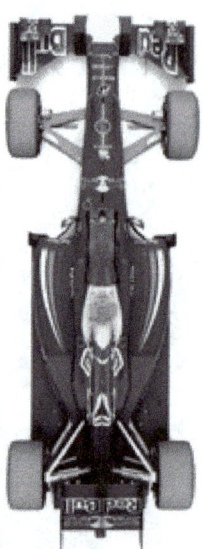

2015
McLaren MP4-30

2014
McLaren MP4-29

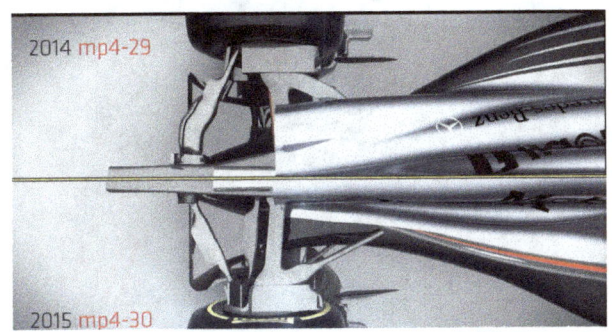

Also in motorcycles:

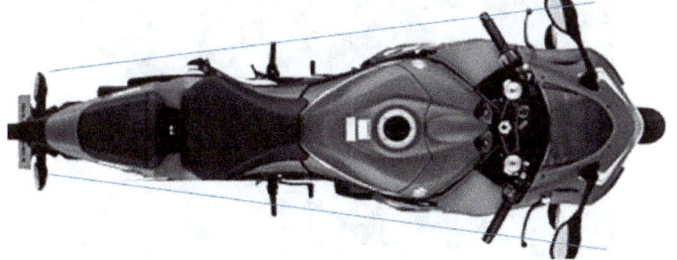

Even the floor adopts a sort of skirt or lips in order to prevent the air from going outside or "falling":

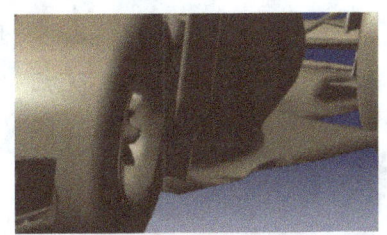

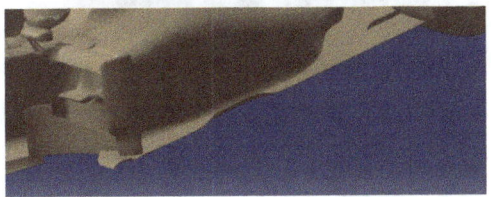

All the efforts are focused on leading the air flow:

The front of the car is also a good area to use Coanda. (we have seen that as a skirt created from exhaust or adding air to up diffuser from exhausts).

In the 2011 season, McLaren used a new design of sidepods, their target was to allow the free movement of air to the diffuser and rear part. In this way, they didn't abuse of the Coanda effect:

We could think of another type of application for the Coanda effect. To control de direction of exhaust fumes, as we can see in the image below, the use of a surface near the exhaust curves the flow:

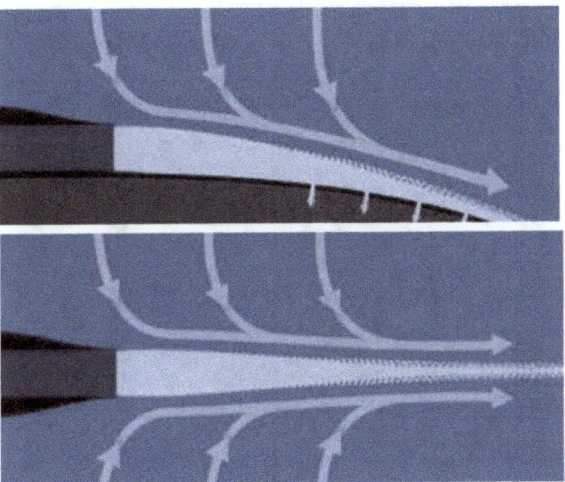

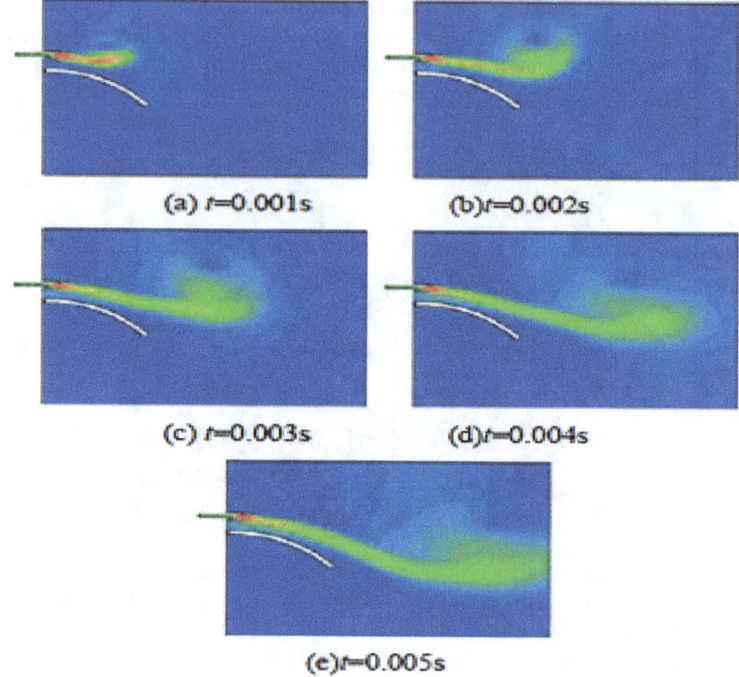

(a) t=0.001s   (b)t=0.002s

(c) t=0.003s   (d)t=0.004s

(e)t=0.005s

    Also, is possible to use one vortex with Coanda, in order to make an air flow where you want to go; under sidepods:

    These produce a vortex which rotation and helped to Coanda, it is directed towards the top of the diffuser.

    Use the exhaust is important, because in the speed, exist depression; this depression may be to use it for deflecting one principal flow to another direction:

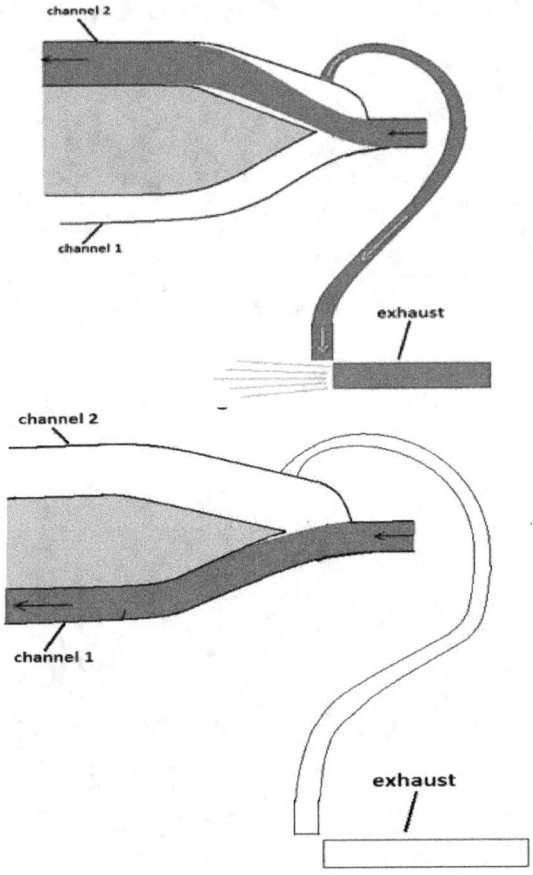

If the rules would allow it, we could use an air injection system on the wings, in order to drag the boundary layer backwards. At the same time, it diverts the flow to the target area:

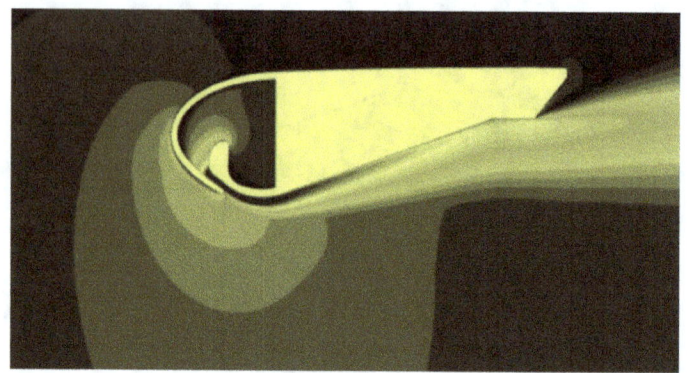

Another application of Coanda effect: about aircrafts, especially in newer models, the exhaust gasses in outlet directional, is very important; that, produce a maneuver very easy:

May be there is a possibility of directing the exhaust gases, from Coanda effect:

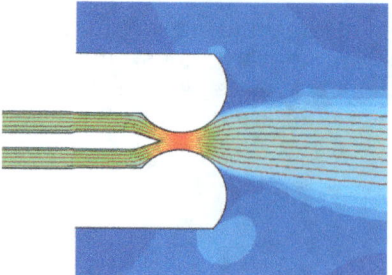

That is; little's changes in geometry (circle red), may be deflect the air flow, go up, for example:

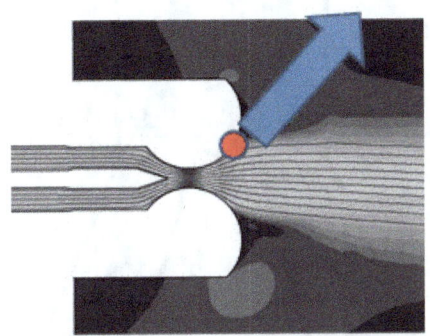

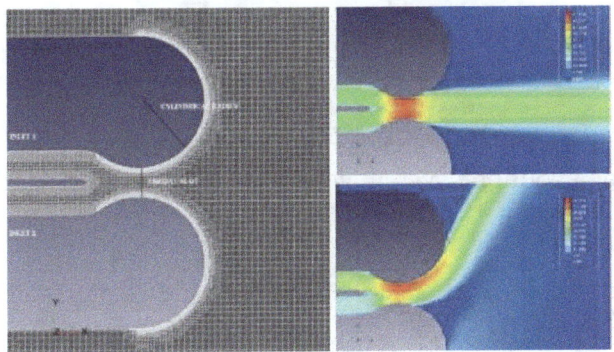

Coanda effect, have a lot sensitivity to small changes iiii

➜ For that, is possible the flow direction CHANGE, if the speed is higher or lower iiii

Also so, the forces and moments iiii

It would be possible to exploit this fact, to design a system to canalize a flow to a place depending on the speed of yaw....

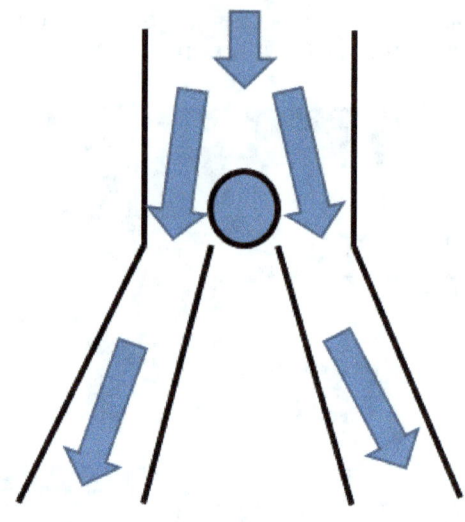

In order to analyze that, we can see the next image about one system with flow vibrating up-down: flow left-right:

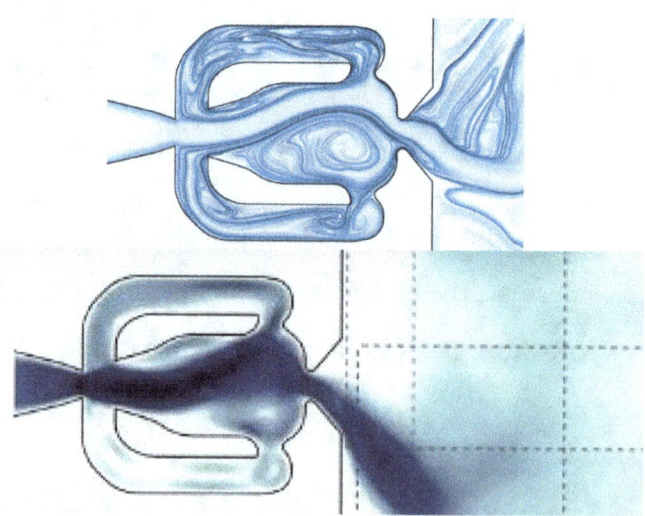

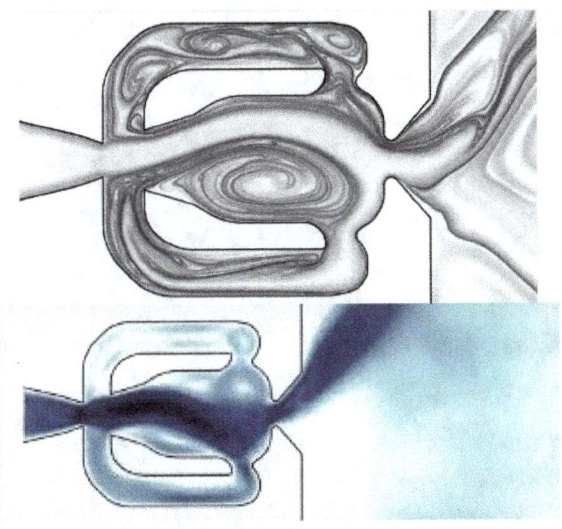

There are also, another system able to adapt the air flow through one direction with little variation of geometry:

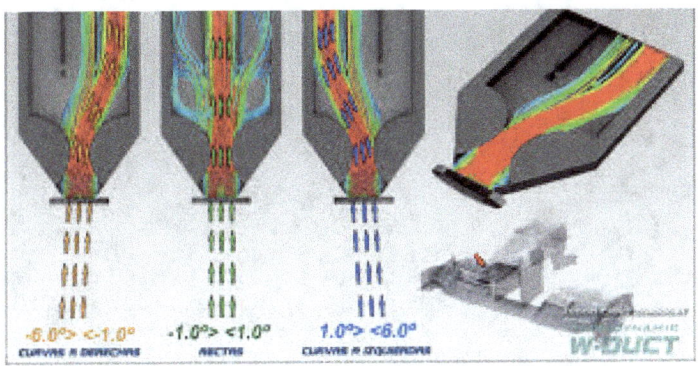

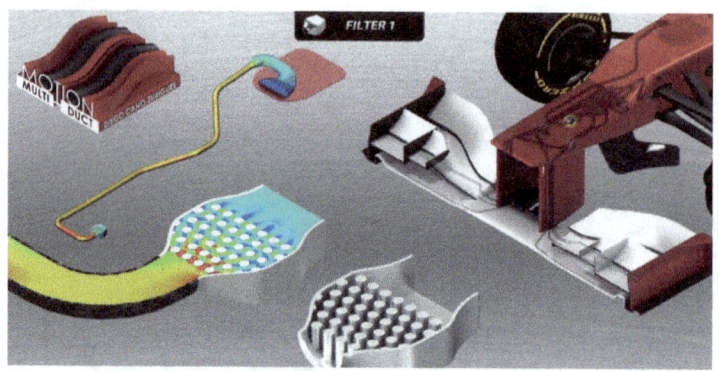

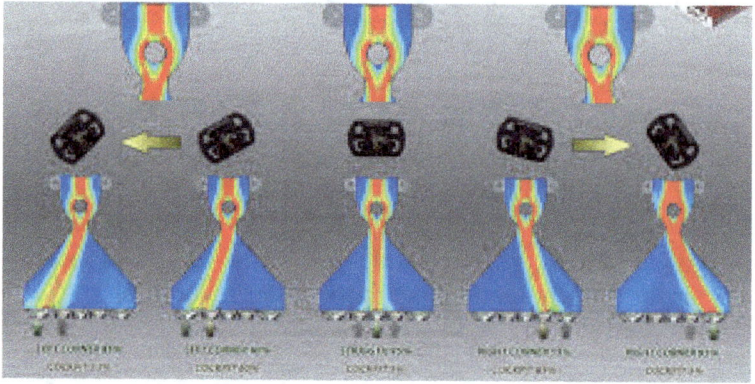

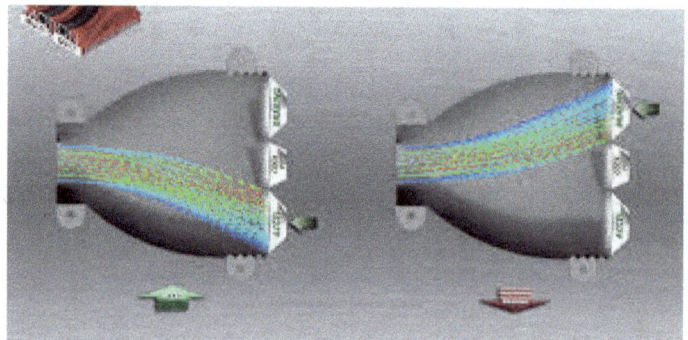

How produce so, these little's variations of geometry "virtual" in order to change the air flow ? Trough holes and low pressure from diffuser or other low pressures zones:

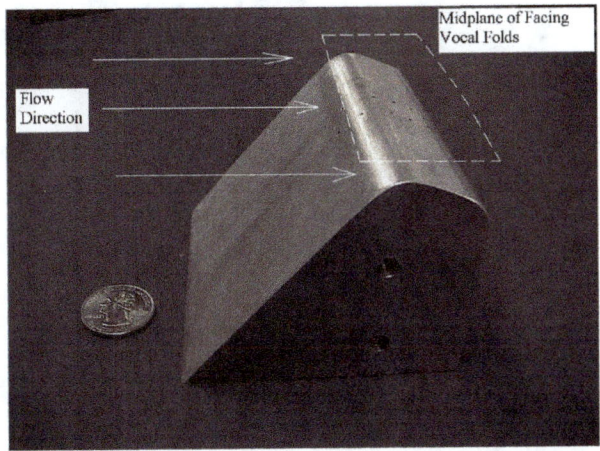

Flow Direction

Midplane of Facing Vocal Folds

→ This toy, match Coanda and Magnus effect: if you tourn the toy, is possible to maintain it in the air:

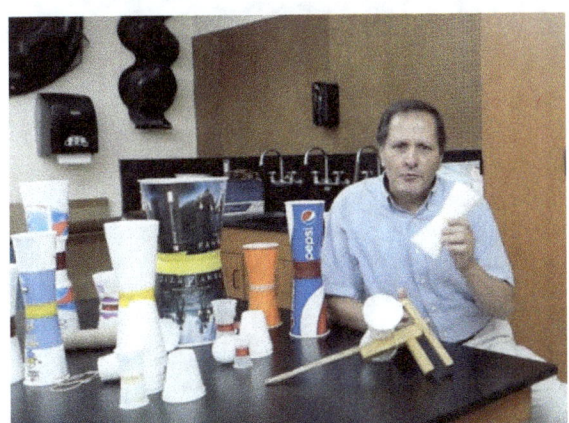